The Physics of Genesis

The Physics of Genesis

Jim Thompson

Pleasant Word
A Division of WINEPRESS PUBLISHING

Pleasant Word (a division of WinePress Publishing, PO Box 428, Enumclaw, WA 98022) functions only as book publisher. As such, the ultimate design, content, editorial accuracy, and views expressed or implied in this work are those of the author.

ISBN 13: 978-1-4141-1160-5
ISBN 10: 1-4141-1160-6
Library of Congress Catalog Card Number: 2007910285

Contents

Preface

This book is written for believers and unbelievers who are not happy with the picture of the world offered by either evolutionists or creationists. At the same time, the people for whom this book is written retain a lively interest in the Bible. They want to know what the text really says but without the usual overlay of denominational agendas. If you are such a person, this book was written for you because it was written by a person just like you.

My journey through Genesis has been a long and winding road. At the age of sixteen I was an assistant to a biology teacher, who for religious reasons refused to teach the theory of evolution. At that time, I had no knowledge of the religious debates surrounding evolution and creation. But I did sense that there was something wrong with his attitude. I wasn't disturbed by the topic of evolution; my concern was different. I had a feeling that this teacher was somehow missing the point of Genesis. That year I stepped in at his request and taught my first class on evolution and Genesis. The point I emphasized then was that Genesis is not so much the opposite of evolution as it is something completely different, something strange and wonderful. That message still stands as the core conviction of this book.

Succeeding years took me through college, a mainstream theological seminary, and finally graduate studies at the University of California, Berkeley. When a person follows that kind of path he or she usually picks up the habit of accommodating the words of the Bible to whatever theory of history or science happens to be in vogue. I was always drawn a different way. I suspected that ancient wisdom offered great insights and that modern people would be better off if we tried to accommodate our thinking to the wisdom of the past, rather than reinterpreting the past in light of contemporary ideas. This instinct was not based on any consciously conservative tendency on my part; it was, rather, based on my observation that the world of the ancients and especially the world of the Bible was a lot more interesting than the rather bland notions being offered by mainstream scholarship then and now. After all, which is the more exciting idea, the claim that God supernaturally created the universe in six short days or the notion that everything just muddled its way into existence through a long slog of mindless accidents? I am not asking at this point which of these ideas is true; I am only asking which idea is more interesting. I suspect that a great many people, nonbelievers as well as believers, would secretly cheer for the claims of Genesis *if they could be somehow presented in a way that makes them credible*.

What stops many people from accepting the claims of Genesis is that they do not find the text believable, which poses a serious problem for them. Because Genesis does not seem credible, such people live in two worlds. On the one hand are all sorts of tales that mom and dad told them such as those about Moses parting the Red Sea and Jesus walking on water, things they might still wish in their heart of hearts were really true. These things they ascribe to the realm of religious faith. On the other hand are all those things they as maturing adults “know” are true, the realm of fact and science. The first world is full of fascinating ideas and inspiring possibilities; the second is redolent of dull lectures in a windowless basement classroom. The sad part is that many of these people presume that the mantle of truth belongs to the droning voice in the basement lecture hall.

This book challenges and reverses that assumption. The reason I can write this book is that my respect for the wisdom of the ancients and my continual technical study of the Bible over many years has confirmed my youthful instinct that we who are troubled by the text of Genesis may not be reading that text accurately. I no longer believe that Genesis is a book of faith. I believe Genesis offers a careful scientific presentation that resembles the best thinking at the cutting edge of contemporary research going on today. In this book I will share my discoveries with you.

Make no mistake; the book you hold is not a creationist tome. Nor is it a text that is friend to any theory of evolution. The ideas you are about to encounter find their correspondence in quantum physics, information theory, and chaos theory. This is so because when we look closely into the words of Genesis we find a text deeply rooted in physics and mathematics. The title of this book is important. This book is not about oddball speculations concerning paleontology, geology, or biology. This is a description and analysis of the *physics* of Genesis that contends that the first chapter of Genesis is a serious presentation that tells us in the language of physics exactly how the universe was made and how it works.

I have not written this book to persuade anyone to believe in the Bible in the religious sense. My objective is to help people better *understand* an important part of the Bible. The findings I report will offer encouragement to the faith of people who want to believe but who find in Genesis, as they understand it, an intellectual barrier to faith. But in the end, faith is always a gift, a gift neither I nor any other author can give. If this book changes the way you read the Bible, I would very much like to hear from you at www.physicsofgenesis.com. I also invite questions or comments that might result from your reading.

This book could not have been written without the help of a multitude of people whose lives touched mine over many years, all of whom deserve thanks. But in this preface I want to extend thanks to the people of the churches I served in my active career as a parish pastor and the people I serve today as a hospice chaplain. It is not fair but, nevertheless, a fact of life that people have to accept as their pastor a person who is, like many of them, struggling to figure things out and for some of us it takes a long time to do that. The good news is that God's grace shines brightest where human weakness and ignorance are most evident.

It is to the people I have served as pastor that I dedicate this book. To them I say, these are the lessons I could not preach when I stood in your pulpits. These are the insights it took a lifetime for me to figure out. I hope a few of you on reading this will sigh and say, "Well, finally he got part of it right!"

I want to express my personal thanks to Jan Holmberg, an extraordinary editor, whose telephone call, out of the blue and at the right time, set my dormant manuscript on the journey toward publication and whose editorial work improved the text beyond description. I also want to thank my wife, Lynn, whose constant pressure to do something with these ideas, pressure I usually took as mere aggravation, actually reflected a deep belief in me and in what I could do.

Above all, to God, who really did in fact create the heavens and the earth in six short days and exactly in the way He said, be all praise and glory, now and forever.

Chapter 1

Radical Scriptures in a Quantum World

Now God worked unusual miracles by the hands of Paul, so that even handkerchiefs or aprons were brought from his body to the sick, and the diseases left them and the evil spirits went out of them.

— Acts 19:11–12

What a strange Bible verse that is! The verse comes from the New Testament book of Acts, the story of the early church, a story that focuses largely on the ministry of St. Paul. An extraordinary man whom God used to bring good news of His grace to the people of the Roman Empire, Paul supported himself during his missionary travels by working as a tentmaker by day and a preacher by night. Laboring in the midday sun, Paul would have worn a sweatband to keep perspiration out of his eyes. This verse is telling us that so much power flowed through Paul's life that people would grab the dirty sweatbands he had worn, bring them to the sick and place them on their skin, and people would be healed of their diseases or released from demonic possession.

Sounds a little nutty, doesn't it? I seem to recall a few television faith healers pitching prayer cloths that sound a lot like Paul's sweatbands. Of course, they didn't offer any sweat and the prayer cloths were never given away (minimum donation required!).

St. Paul's sweatbands are not the only oddity we find in the Bible. One of my favorite passages involves the prophet Elisha, an eighth century BCE miracle worker active in the northern part of Israel. Stories about him are told in the second book of Kings in the Old Testament. One of those stories goes like this:

> And Elisha returned to Gilgal, and there was a famine in the land. Now the sons of the prophets were sitting before him; and he said to his servant, "Put on the large pot, and boil stew for the sons of the prophets." So one went out into the field to gather herbs, and found a wild vine, and gathered from it a lapful of wild gourds, and came and sliced them into the pot of stew, though they did not know what they were. Then they served it to the men to eat. Now it happened, as they were eating the stew, that they cried out and said, "Man of God, there is death in the pot!" And they could not eat it. So he said, "Then bring some flour." And he put it into the pot, and said, "Serve it to the people, that they may eat." And there was nothing harmful in the pot.
>
> —2 Kings 4:38–41

I like that story perhaps because the reaction of the men reminds me of what happens when I serve my meatloaf to my family. I can even relate to the strategy of putting flour into the meal, in my case in a futile effort to thicken the already cooked gravy. (Did you know you get a sort of beefy Elmer's glue when you do that?)

Passages of this kind have gone a long way toward putting some people off from listening to the message of the Bible. I was one of those people. But now I am inclined to a different view. I tend to think of passages of that kind as the most interesting and important material in the whole Book.

Preventative Thinking

When people open a Bible, what do they hope to encounter? It seems to me that most believers and nonbelievers as well would like to find themselves addressed by God. But what does that mean? Again, it seems to me that most people want to be told something they did not know before. A lot of readers would be disappointed if the Bible offered nothing more than commonsense observations and textbook orthodoxies.

But it is at this point where many people let themselves down. We often ignore passages of the kind I quoted, in effect reading them out of the Bible, and restrict our attention to more commonplace passages and pedestrian interpretations that offer no challenge to our habitual ways of seeing the world. When we do that the Bible becomes pedantic and we become disillusioned. We came to the Bible hoping to hear a word from the Lord but all we get is a reiteration of things we have been told all our lives.

That is precisely the reason passages of the kind I quoted become important. These passages, verses that some call "radical scriptures," are the amplifiers that open our ears to hear the deeper message of the Bible. They hint that more is going on in this book than we may have guessed from the sermons we may have endured as children. Radical scriptures reveal a view of the world unlike anything we learned in public school or even in most Sunday schools.

Any satisfying reading of the Bible will begin, I believe, with careful attention to those parts that qualify as radical scriptures. I say that because only such a reading will give us new information that is both interesting and potentially enlightening. This does not mean that we turn our back on everything we know and believe and instead just read the Bible. It does mean that we let the Bible teach us how to approach our world with a sense of wonder again. Instead of starting with conventional wisdom and squeezing the Bible into that narrow frame, it is a lot more fun to start with the Bible and let God's Word challenge conventional wisdom.

There is a good reason for doing this. Peril, whether for an individual or a civilization, arises not from the things we recognize that we do not know but rather from things we think we know but that are not so. Ignorance, when it is perceived as such, should not frighten us because ignorance is an emptiness that will be filled. Knowledge, on the other hand, is a container already filled. If the container is filled with truth, well and good. But if the container is filled with illusion, then we have a problem. Ignorance requires only instruction, but deceptive knowledge demands a deeper fix. When knowledge is false, our minds must first be emptied of errant premises and faulty assumptions before any learning can take place.

Think about this radical scripture spoken by Jesus Christ, "I say to you, if you have faith as a mustard seed, you will say to this mountain, 'Move from here to there,' and it will move; and nothing will be impossible for you" (Matt. 17:20). Many accept this verse as a piece of religious inspiration, hyperbolic perhaps, but well intentioned. I am suggesting that we try accepting Jesus' words as a statement of fact. The inspirational approach turns the Bible into an extended greeting card, something sweet but insubstantial. But when we accept Jesus' words as a statement about reality, those words open a series of questions that have the advantage that they are, at the very least, interesting.

Is the physical appearance of this world something fixed and beyond our control or is that appearance in some way related to the way we observe our world? It might surprise some people to know that there is an entire field of physics, quantum mechanics, that is based partly on the *law of nonseparability*. Stated briefly (more on this later) nonseparability means that our observations play a role in giving our world shape to such a degree that we cannot speak of the material world as if it were something separate that possesses its own existence apart from the way we observe and measure it. In light of quantum physics it turns out that Jesus' statement about moving mountains by the power of transformed perceptions finds an echo in contemporary research. The problem was that it took us two thousand years to discover the scientific principles underlying Jesus' assertion. We might have a deeper understanding of our universe today had we accepted Jesus' statement even provisionally as a description of the operation of our world and asked ourselves, "If Jesus was telling us the truth, what kind of universe might we be inhabiting?"

Radical scriptures open our minds to unimagined possibilities. In the case of Jesus' words, many believed in the messenger but few actually investigated His message, at least on this point. As such we fell into the most dangerous type of error. Blaise Pascal defined two types of mental errors. A type I error rejects the truth of a hypothesis outright. A type II error accepts a false hypothesis. Of these errors, the most dangerous, Pascal said, is the type I error. Experiment and honest research will soon expose a false hypothesis. But with the type I error, a possibility is never even explored. The greatest damage done to the lives of persons and civilizations is often done by the paths not trodden, the risks not taken, and the ideas deemed too radical to carefully examine. Many cultures have followed the ruts of ingrained habits and unquestioned customs into environmental, political, or social catastrophe. Radical scripture passages offer a preventative against deadly type I errors by challenging every settled assumption about our world. But to use these scriptures in that way we need to follow a few rules.

I recall a radio talk show where a caller complained about various criticisms directed against Islam. Arguing that Judaism is just as brutal and nasty as any Islamofascist terrorist organization, the caller pointed to the story about the golden calf in Exodus 32. In that story, the people made a golden calf and worshiped it, saying, "This is your god!" (v. 4). The response was violent. Moses gathered to his side those loyal to the God of Israel and told them, "Thus says the Lord God of Israel: 'Let every man put his sword on his side, and go in and out from entrance to entrance throughout the camp, and let every man kill his brother, every man his companion, and every man his neighbor'" (v. 27). The Bible tells us that about "3,000 men of the people died that day" (v. 28).

The caller wanted his audience to understand this event as an example of religious intolerance. But to make that point, he had to ignore the main thrust of the text. The events in Exodus occur in the very presence of God. This is not a group of people getting all riled up at a church service. These people are seeing the flames of God's power, they are wrapped in His clouds, and they are literally hearing God's voice in the thunder. This is Israel at Mt. Sinai. On top of that, the people had just made a solemn pledge to God, a pledge spoken face to face: "All that the Lord has spoken we will do" (Ex. 19:8). The actual presence of God together with the sacred vow so fresh on their lips changed their construction of a pagan idol from a lifestyle choice into an act of suicidal insanity.

Taking into consideration the context of the event, there is no parallel whatsoever between the three thousand who died that day at Sinai and the three thousand who died in the collapse of the Twin Towers. But the caller's comment underscores a typical problem in contemporary biblical interpretation, a tendency on the part of many to read the supernatural elements out of biblical events and so reduce the texts to a matter of moral instruction.

Biblical Reality

Any accurate reading of a biblical text has to include the supernatural element that will always be part of the context of the text. It won't do to remember what the people said and did while forgetting the setting within which they did it. But that is precisely what many readers do. They forget that the worldview of the Bible is radically different from any secular-progressive worldview. Secular worldviews presume a closed universe where human actors dominate the scene and God is completely absent. In the biblical worldview, God is always the most important actor on stage in every circumstance and every event.

This means that the Bible will be a difficult book for secular people to read. Because we have all been influenced by secularist presumptions, it will take hard work (and not a little Divine help) for any of us to move out of our habitual ways of thinking and extend ourselves any distance into the biblical thought world. But I believe we will find that effort worthwhile because of the fresh insights and useful information we will gain.

The information I speak of is not the same as believing. As I said before, in this book I am neither seeking to advocate faith nor trying to trick someone into becoming a believer. I am speaking about accuracy. We cannot gain anything from the Bible unless we read it accurately. Once we have read the Bible we can each decide if we believe what it says. But we cannot read the Bible accurately unless we approach it with serious attention to the actual content of the text and that content necessarily includes the supernatural context within which every Bible story is framed.

In the story about the golden calf, the text has nothing to do with religious tolerance or intolerance. But read in its supernatural context it does say something about the human heart. The point of the story is that even if we stood in the very presence of God with every blessed assurance of that presence exploding in front of us, we human beings would still rebel against Him. Many say that if only God would show Himself to them in miracle, sign, or wonder then they would believe. The story about the golden calf answers and says, "No, you would not believe." The story tells us that human ways of understanding are not just uninformed; they are malignant, infected with a cancer that keeps our minds from ever functioning as they should. This text calls into question the accuracy of all our premises and all our assumptions; it questions even the truth of our observations. When it comes to understanding reality, we are, it says, infected with Pascal's type I error. Our minds are clouded not by false ideas but by our tendency to deny the truth that stands directly in front of us. This is especially the case when we are confronted with insights that arise from a source beyond our world. "'My thoughts are not your thoughts, nor are your ways My ways,' says the Lord" (Isa. 55:8). Human habits and thoughts are not just different from God's ways; the Bible reveals that they are usually the very opposite of God's ways and therefore tend to be not just a bit misguided but fundamentally wrong and dangerous.

If we are going to let the Bible help us break out of our illusions we have to take its supernatural context seriously. But there is another condition that must be met. We must also take the text literally.

One of the worst things we can do to the Bible is to spiritualize its message. Many people routinely take the plain text of the Bible and try to apply interpretations to it that have nothing to do with the literal content of the text. The most common example has to do with Israel. It is commonplace in many denominations to read the Hebrew Bible (the Old Testament) in such a way as to refer all statements about the future of Israel to the Christian church. The problem is that the Bible is packed with prophecies that are specific and precise and related to the land and nation of Israel. They offer a vision of a future for God's chosen people that has not yet found fulfillment but stands as a source of hope not only for Israel but for the whole creation and every person inhabiting it. This hope-filled vision is lost when promises made to Israel are squeezed into the rather small box that is the Christian church. When that happens, texts that offer a cosmic vision of a renewed planet get confined to the dimensions of a church council meeting and the church itself loses sight of its place in a much larger picture.

The Bible is interesting only to the degree that we take it literally. A literal reading of the Bible informs and instructs us. But when we turn its wild promises into symbols and metaphors that refer to nothing greater than things we are already doing, we turn a tiger into a pussycat. Our hearts, which yearn to be addressed by a Word that comes from God, a mighty Word that challenges our premises and assumptions, are left hungry.

Look with me at what can happen if we begin to read the Bible with respect to the literal meaning of the words. I want to share with you a passage from the Bible that will prove to be either one of the murkiest bits of writing you have ever seen or the most astonishing:

> A period of seventy sets of seven has been decreed for your people and your holy city to finish their rebellion, to put an end to their sin, to atone for their guilt, to bring in everlasting righteousness to confirm the prophetic vision, and to anoint the Most Holy Place. Now listen and understand! Seven sets of seven plus sixty-two sets of seven will pass from the time the command is given to rebuild Jerusalem until a ruler, the Anointed One, comes. Jerusalem will be rebuilt with streets and strong defenses, despite the perilous times. After this period of sixty-two sets of seven, the Anointed One will be killed, appearing to have accomplished nothing, and a ruler will arise whose armies will destroy the city and the Temple. The end will come with a flood, and war and its miseries are decreed from that time to the very end. The ruler will

> make a treaty with the people for a period of one set of seven, but after half this time, he will put an end to the sacrifices and offerings. And as a climax to all his terrible deeds, he will set up a sacrilegious object that causes desecration, until the fate decreed for this defiler is finally poured out on him.
>
> —Dan. 9:24–27

As we read this passage one thing becomes immediately clear. These words resist metaphorical interpretation. The language is specific and precise. Daniel was a Jew, so the words "your people" are easily understood. The phrase "your holy city" can only refer to Jerusalem, and, indeed, Jerusalem is named in the text. It is clear that after a set period of time someone called "the Anointed One" will be killed and the holy city will be destroyed along with its temple. There is much more we could say about this text but that would be the subject of a different book. The question before us is what happens when we read the Bible with attention to its literal meaning instead of spiritualizing it.

When we read the text literally what catches our attention are the numbers. It is easy to see that the text offers predictions concerning events that were future to the writer. The unusual feature of these predictions lies in their link to specific numbers. What do the numbers mean?

A Book of Numbers

The first reference to time speaks of "a period of seventy sets of seven." To understand that phrase it is important to know a bit about the background behind these verses. In 606 BCE, many prominent Jews in Jerusalem were taken captive and forced into exile by the Babylonian Empire. Daniel, the man who reported this prophecy, was one of those Jews. Daniel, now an elderly man living in Babylon, had been reading the words of the prophet Jeremiah, who long before it happened had predicted this exile and said that it would last seventy years. Seeing that this time was coming to an end, Daniel cried out to God, asking for a revelation regarding the future of his people. In response to his prayer, he tells us (v. 21) that he was visited by the angel Gabriel who communicated the message we just read. The substance of the message was that God will not abandon his people. He had, in fact, a detailed plan for the future of Israel. Gabriel described the key events in this plan, but more than that, fixed those events to a precise time frame.

How precise was this time frame? The context makes clear that the reference to "seventy sets of seven" refers to years. The issue that troubled Daniel was what will happen after the seventy *years* of exile. The length of the years in question was defined by the physical setting of the book. The book of Daniel was written in Babylon, where the sacred calendar consisted of a 360-day year. It is noteworthy that the same 360-day sacred year was observed by Jews, Egyptians, Mayans in Central America, and many other people scattered all over the world.

The text speaks of two time segments. The first segment consists of "seven sets of seven plus sixty-two sets of seven" (v. 25) or a total of sixty-nine sets of seven. It is helpful to translate this into days. Using the Babylonian and Jewish sacred calendar, the days in question here total 173,880 days. This time segment begins at a very specific point. The starting point is "the time the command is given to rebuild Jerusalem" (v. 25). Do we know the date on which that command was given? The answer is that we do. Persian records tell us that authority to rebuild the city of Jerusalem was given by Artaxerxes Longimanus on March 14, 445 BCE.

The clock on Daniel's prophecy began ticking on that date. Now the question becomes, what happens when the clock runs out? According to the text, "a ruler, the Anointed One, will come." Since the entire prophecy has to do with the fate of Jerusalem, the words "will come" can only mean come to Jerusalem. The words "Anointed One" are equally clear. These words always refer to the long-awaited Messiah of Israel, the Son of David whose Hebrew name means "anointed," which, translated into Greek, is rendered as "Christ."

So let's do the math. We need to add 173,880 days to the starting date of March 14, 445 BCE. What do we get? We get the date April

6, 32 CE What is significant about that date? April 6, 32 CE falls exactly seven days before Passover eve that year, the day on which Jesus would be killed. The end date of the prophecy was the day that would come to be celebrated in the church as Palm Sunday. This was the day Jesus Christ entered the holy city of Jerusalem with the pomp and honors due a king, honors given not by the city as a whole but only by Jesus' immediate disciples. It turns out that Daniel was given the exact date on which Israel's Messiah would enter Jerusalem as king more than five hundred years before He arrived.

There is, obviously, much more that could be said about the prophecy in Daniel. But this is not a book about Daniel or about prophecy. The point I want to make has to do with the specificity of biblical language. We could read this entire passage as a set of symbols and metaphors and come away puzzled and even a bit bored by the whole thing. Or we could read the passage literally and in doing so encounter something astonishing. This book chooses the latter course.

Unfortunately, such attention to literal detail is not the usual way in which the Bible is read. The events of Holy Week culminating in Israel's execution of her own Messiah make it clear that Israel had stopped reading its scriptures with careful respect for their literal meaning. It is precisely this failure that Jesus spoke of as He approached Jerusalem and looked upon a city where most people were unprepared to receive their King and said, "If thou hadst known, even thou *at least in this thy day*, things which belong unto thy peace! But now they are hid from thine eyes (Luke 19:42, emphasis added).

The key words in that passage are contained in the phrase "even thou at least in this thy day." The Bible had revealed to Israel not just the plan for her future but the very day on which a key event, the visitation of her Messiah to His temple, would happen. Because the nation had forgotten to attend to the details of prophecy, Israel had lost the opportunity for the immediate fulfillment of her hopes. The results were destined to be tragic. Jesus' lament concludes with a prophecy of His own:

> Days will come upon you when your enemies will build an embankment around you, surround you and close you in on every side, and level you, and your children within you, to the ground; and they will not leave in you one stone upon another, *because you did not know the time of your visitation.*
>
> —Luke 19:43, emphasis added

Note the insistence that it was because Israel did not know the time of her King's coming that destruction fell upon the city and the nation was destroyed. Israel had been told when its Messiah would come right down to the day it would happen but because they did not give attention to the details of the text most of the nation was not ready to receive Him. True to Jesus' prediction, thirty-eight years later the fifth, tenth, twelfth, and fifteenth Roman legions laid siege to Jerusalem. In an effort to find treasure purportedly hidden there, the Roman army accidentally burned the temple. The soldiers were then ordered to search every corner of the scorched city for melted gold until not a "single stone" remained in place and the city as a whole was leveled.

I have offered this excursis to make just one point. The Bible challenges us to read its text with attention to the literal meaning of its words. This is particularly so in regard to radical scriptures of the kind we have just examined. Those texts most outrageous in their claims are exactly the sort of text that require closest attention to their literal content. In fact, only in the literal meaning of the words does the message come through at all.

This does not mean the Bible doesn't use figures of speech to render its message. Bible scholars have identified more than two hundred different kinds of figures of speech at work in the text. A literal reading of the Bible requires that a figure of speech be taken for what it is. If a figure is being used, then a figurative interpretation is the literal interpretation. But if the text presents a matter as historical it must be read that way even if the text seems to be in error in light of current historical knowledge. We cannot receive anything new from the Bible unless we first let it speak in its own voice before we as readers react to what it says. We will gain nothing if we let our premises and assumptions predetermine what a text can say but we might learn much if we permit a text to challenge our presuppositions.

We will turn now toward one of the strangest examples of what we are calling a radical scripture to see what happens when we read it in just that way. The scripture I speak of is Genesis 1, the story of the creation of the world.

Chapter 2

Clearing the Ground

Radical scriptures come in two flavors. There are those that deal with matters germane mostly to biblical interpretation. In these cases the challenge is simple: read the text and pay attention to what it says. In the example of the dating of the arrival of the Messiah, all we had to do was read the text and count the days and the message delivered itself. But another kind of radical text contradicts ingrained habits of thinking ensconced in a given culture. Genesis 1 falls into this category.

Whenever we encounter texts that challenge cultural conventions, it is not sufficient to simply read the text. The problem is that we cannot receive new information when our minds are already full. This problem is exacerbated when the habit of mind is supported by an ideological commitment. In that case, evidence alone will not be enough to overcome the errant idea. Evidence informs the mind but dogma lives in the emotions. All the evidence in the world cannot overcome the stranglehold of a falsehood that has become a dogmatic commitment.

A good example of this problem is the dominance enjoyed for centuries by a cosmology that placed the earth at the center of the solar system. There was always evidence that this cosmology did not quite work but it was not evidence that held the theory in place.

The geocentric picture of the solar system was sustained in some measure by fear of the punishment that would be imposed by the Roman Catholic Church upon anyone challenging this picture. More powerfully, however, it was held in place by a mutual consent born of fear of what would happen to a culture deprived of the only cosmology it knew.

Today we find ourselves in exactly the same situation regarding the theory of evolution and various theories of origin associated with that idea. As we will demonstrate, these notions have no weight as scientific theories but they are sustained by deeply felt emotional commitments shared by their supporters. For many, these are the ideas that hold their world together, a world based on faith in what they think of as "science." Even if an idea has no merit, the difficulty is that until the emotional attachments that support it are overcome it is hard for science or the culture as a whole to embrace more likely possibilities. It is also impossible under such conditions for a wide audience to glean any insights from the literal text of Genesis no matter how true or helpful such insights might turn out to be. Although Genesis clearly offers a compelling picture of how things work, it is human nature to defend our habits, mental as well as behavioral, even if those habits are based on beliefs that are completely wrong. In this chapter I will present information that can help us move beyond the speculations stimulated by the writings of Charles Darwin. I do not intend to fully analyze or refute Darwinian conjectures (others have already done that very successfully). I simply want to establish that Darwinism is not the collection of irrefutable facts that many nonscientists assume it is and in so doing prepare us to take a fresh look at questions of origin.

At the level of evidence, the Darwinian theory of evolution can quickly be dispatched.

Specious Speculation

When a scientist offers a hypothesis, s/he must also offer predictive statements based on the theory that allows it to be falsified depending on how the predictions turn out. As a scientist, Charles Darwin offered such statements.

First let's look at what Darwin proposed. Darwin speculated that all life forms can be traced to a common ancestor and that present life forms have emerged through the gradual and steady accumulation of incremental changes. All this occurred under the influence of natural selection, a process that would, he thought, take the form of a tendency on the part of creatures possessing a survival advantage to generate offspring who would live and propagate themselves in greater numbers than those lacking the advantage.

To shape this speculation into a hypothesis, Darwin offered three predictions that presented opportunities for testing that could falsify the theory. Darwin's first test concerned the fossil record. He said: "The number of intermediate and transitional links between a living and extinct species must have been inconceivably great."[1]

If Darwin's theory was correct, the natural world would have a unique look. It would be a world in which taxonomic identification would be exceedingly difficult because biotic communities would exhibit constant morphological change. Scientists would not search for missing links because no form would achieve such stable expression as to suggest the need for a link to another form. In other words, the question of intermediate forms would not come up because nature would offer nothing but intermediate forms.

Darwin was aware that in his day the fossil record did not support his theory, but he proposed that with time it would. Unfortunately, the evidence he hoped for has not appeared. Paleontologist Niles Eldridge has reviewed the fossil record as it is known today and has stated its evidence in this way:

> Evolution, at least on a grand scale is not forever tinkering, trying to come up with a better mousetrap. It's the other way around: species, and the ecosystems that their component organisms staff, are tenacious. They "work" perfectly well, and, once entrenched, are unlikely either to change or to be displaced by newly evolved taxa—unless and until extinction knocks ecosystems off their tracks. Then evolution breaks loose. [2]

Eldridge claims that the most prominent characteristic evidenced by nature is not change but stability. Referring to the life work of a prominent paleontologist, he writes:

> Lieberman's…fossils came from an ecological setting—the Middle Devonian Hamilton Group of eastern North America–in which some three hundred species have been discovered. The vast majority of those species are present at the beginning of Hamilton times. Most make it to the very end–when environmental conditions changed and most species became extinct. Many show some signs of change as the six million years went slowly by. But the last members of nearly all those species ended up looking very much like their forerunners six million years earlier. It is a stunning fact that it is possible to illustrate the variation within each of these species in the same small number of photographs or drawings used to depict variation (males, females, juveniles; seasonal plumage changes; geographic variation) in a typical modern guidebook to today's birds in eastern North America. This is stunning because it means that six million years of evolutionary history rarely increased the variation much beyond what was already present within a species over its entire geographic range at any one point in time. This is not what Darwinians of any stripe would expect.[3]

The fossil record reveals a pattern of persistence. Species survive as identifiable units for millions of years until, in the end, they usually become extinct. The following sentence needs to be clearly understood. After 150 years of searching for "missing links," the fossil record does not offer one single example of a species evolving into another kind of creature. Changes that are observed usually turn out to be a matter of oscillation as different groups of genes that were always part of the genome of a particular species achieve dominant expression under changing conditions. Eldridge described the process in this way: "Anatomical traits do shift around a bit as time goes by. But rarely do we see progressive transformation in any one direction lasting very long. What we see instead is oscillation. Variable traits usually seem to dance around an average value."[4]

The oscillations that we actually observe in nature (the emergence of pesticide-resistant insects, for example) have nothing whatsoever to do with the emergence of a new species. Quite the contrary, these sorts of changes tend to assure the survival of an existing species. Darwinism is a speculation about an event that has never, not even once, been observed in either nature or a laboratory.

Can we consider Darwin's theory to be falsified on this particular point? Darwin's supporters widely admit that the fossil record does not support their speculation. But they argue that the record is still incomplete and that when it is complete it will sustain their position. What this argument does not address is the proven stability of the natural world over most of its history. Carl Sagan used to say that critics of Darwinism had to remember that nature had billions of years in which to shape the biotic community we see. His argument was hard to refute because given enough time almost any kind of development would seem to be theoretically possible. The problem is that nature seems to waste most of its time. We simply do not see the gradual and steady morphological changes that would make those billions of years a useful explanation for evolution. It may well be that given enough time and enough monkeys one could generate a folio of Shakespeare but it won't ever happen no matter how many monkeys you have or how long you give them if they will not type. It appears that nature, for the most part, simply does not type. Nature wallows in lazy stability for millions of years and only becomes active in sudden bursts connected with extinction events.

Natural Leaps

This brings us to the second falsification test that Darwin offered along with his speculation. Darwin's idea requires that natural changes be gradual and steady. This is so because evolution as he defined it was driven by only one factor, the natural tendency of organisms bearing a beneficial adaptation to thrive and reproduce in greater numbers than creatures lacking that adaptation. If nature exhibited sudden leaps that resulted in large-scale changes in biotic communities, such leaps would indicate that some other force besides natural selection was at work and research would have to be redirected toward identifying that mysterious force. Such a study would necessarily draw biology away from Darwin's single-minded focus on natural selection by suggesting the possibility that other factors might play an even more significant role in natural history. Acceptance of the possibility that nature could exhibit sudden leaps might even lead to a reexamination of the possibility that natural history could be influenced by the actions of a Creator. With these possibilities in mind, Darwin had to admit that any example of a leap or "saltation" found in the history of natural life "may be truly urged as a valid argument against the views here entertained."[5]

From the very beginning even Darwin's supporters sensed trouble in this falsification test. Before the *Origin of Species* was published, Darwin's friend T.H. Huxley privately wrote to him, "You have loaded yourself with unnecessary difficulty in adopting natura non facit saltum [nature never makes leaps] so unreservedly."[6] Another friend, geologist Charles Lyell, quickly distanced himself from Darwin's work and pointed to the likelihood that forces other than natural selection were at work in nature with the comment, "I feel Darwin and Huxley deify secondary causes too much. They think they have got farther into the domain of the unknowable than they have by the aid of variation and natural selection."[7]

The question is, does nature make leaps? If it does, the theory as presented by its adherents has been falsified at a foundational point. One of Darwinism's most vigorous advocates, Richard Dawkins, underscored the importance of the saltation test and confirmed that saltations would throw the door open to old approaches to natural history:

> In the context of the fight against Creationism, gradualism is more or less synonymous with evolution itself. If you throw out gradualness, you throw out the very thing that makes evolution more probable than creation. Creation is a special case of saltation—the saltus is the large leap from nothing to fully formed modern life. When you think of what Darwin was

> fighting against, is it any wonder that he continually returned to the theme of slow, gradual, step-by-step change?[8]

The problem for Darwinism lies in the fact that nature most certainly does make leaps. Today every high school student knows that the most important event that paved the way for the emergence of mammals as the dominant species on earth was not the gradual accretion of inherited characteristics. The event that changed the world was almost certainly a rock falling out of the sky. In a single instant, billions of beautifully adapted creatures were obliterated and unimaginable tons of biomass were fated for extinction.

Geologists tell us that this was no isolated event. Similar mass extinctions have been noted at the end of the Permian Age (97 percent of the biomass of the world destroyed) and at the beginning of the Cambrian Age. The latter event offered an especially vexing problem for Darwin himself. Often called biology's "big bang," Cambrian fossils (530 million years old) record the abrupt appearance of most contemporary phyla (basic morphological types). In plain language what this means is that there is absolutely no fossil record demonstrating any kind of evolutionary history connecting the major body forms (phyla) that exist in our world today with each other or with a common ancestor. This gap in the evolutionary chain is more than unfortunate, it is devastating. When evolutionists spin out their scenarios about the history of the various types of creatures that inhabit our world, what they are offering is nothing more than pure guesswork when it comes to the most important part of that history, the emergence of the major phyla.

If nature makes mysterious leaps, no single speculation can ever claim to fully describe the operations of the natural world, especially if that speculation can only account for incremental changes. In light of this fact, while anyone is free to maintain a bias in favor of Darwinism and its insistence that the history of life is entirely a matter of slow, steady processes of accretion driven exclusively by natural selection, that bias can only be a matter of personal choice, a highly questionable choice given the evidence in hand but a choice nevertheless. The problem is that a choice, by definition, is a preference not a theory.

Darwin offered a third test by which his proposal could be falsified. Darwin stated that unless some mechanism could be identified by which acquired characteristics could be inherited by offspring, his speculation would have to be abandoned.

Darwinism was a product of some pretty unsophisticated science. It was, after all, based on scientific knowledge current in the nineteenth century. Darwin knew nothing about the operation of life at the cellular level, operations that are familiar today to any high school student. He was content to observe morphological changes and muse about possible reasons such changes might have occurred.

Darwinism has always been strong on explanations. It can tell us "why" a trait might have developed, but it has always had a problem developing testable ideas that might tell us how that trait might have found its way into the ongoing life of a species. Inevitably, Darwinism took the shape of the kind of children's tales Rudyard Kipling popularized, fables he called "Just So Stories." Darwinists spun colorful narratives that purported to explain why the giraffe got its long neck or the elephant got its trunk (it was always "just so" a creature could do this or that). The problem was that explanations are not causes nor do they offer falsifiable predictions, which means they really are just fables rather than testable scientific proposals.

In particular, Darwin and his first disciples had no idea how an advantageous adaptation might actually be passed along to a creature's descendants. Darwin's speculation was a typical nineteenth-century mechanistic proposal, except in Darwin's case the machine had no working parts. Because of this failure, after causing an initial sensation Darwinism quickly disappeared from public conversation.

Darwinism was salvaged by the rediscovery of the work of Gregor Mendel. The rescue of Mendel's work from years of neglect led to the founding of the science of genetics. Darwinists seized on selected aspects of genetic science to provide themselves with something that, at least at first, appeared to provide an actual mechanism that could give a measure of credibility to their proposal.

Unnatural Mutation

The vehicle that drove evolution under their reworked proposal (which came to bear the name "neo-Darwinism") was a process geneticists call "random genetic mutations." The idea is that every so often biological information changes. These changes are entirely unpredictable, they move toward no purpose, and they are guided by no drive toward meaning or progress. These changes involve nothing more than casual errors in the transcription of the information coded in DNA. Darwinists proposed that when such changes generated an immediate benefit for the creature involved and when the changes could be inherited, such mutations would be favored by natural selection, which means that the creature would survive and produce more offspring that would inherit the mutation and would pass it on to their descendants. Through the accumulation of such random transcription errors (using here an example actually favored by Darwinists), an aquatic bear would eventually become a whale.

It is important to understand that the credibility of Darwinism today rests entirely upon the capacity of mutations to do all the things Darwinists need them to do. Consider what Burton Guttman wrote in his 1999 textbook, one of the two primary textbooks published by the McGraw-Hill Company for use in university biology classes throughout the United States: "Mutation is ultimately the source of all genetic variation and therefore the foundation of evolution."[9] Note also this statement from the textbook prepared by Peter Raven and George Johnson, the other major textbook offered by McGraw-Hill for use in university biology classes: "All evolution begins with alteration in the genetic message. ...Genetic change through mutation and recombination provides the raw material for evolution."[10]

If mutations cannot provide the engine for evolutionary change, then Darwin's proposal still lacks a credible mechanism and must, once again, be relegated to the far fringes of scientific debate. Unfortunately for Darwinists, mutations have not proven to be the blessing they appeared to be when neo-Darwinism was cobbled together back in the 1930s.

The biggest problem involving mutations is that there simply isn't enough time for mutations to do what Darwinists ask them to do. Initial enthusiasm for the efficacy of mutations in evolution was connected to the doctrine of geological uniformitarianism. This is simply a fancy name for the assumption that there are "billions and billions of years" available to account for the evolution of life, enough time to lend credibility to almost any speculation about how life developed. As we have seen, research has not been kind to this idea. It turns out that we don't have anywhere near the billions of years required for mutations to work their evolutionary magic because change in biotic community seems to be related to extinction events. At this point it appears that when an old biotic community is eliminated a different biotic community emerges and assumes a stable pattern. Scientists tell us that this usually happens within ten million years at the most. But this "ten million year" span is nothing more than the shortest time frame we can measure when speaking of ages so ancient to us. The actual emergence of the new community could easily have been much faster and many scientists speculate that it probably was.

This poses a difficulty for neo-Darwinism because while the exact number of years in which these quick bursts of transformation took place is not known, the standard rate of mutations in the genetic material of living cells is well known. This mutation rate can be measured in mitochondrial DNA and in the DNA of Y chromosomes (whose information changes very slowly over time except for those changes generated by normal transcription errors). Using the standard rate of mutations it is clear that the span of time it would take to develop a whale out of an aquatic bear, as some Darwinists propose, would encompass an age longer than the estimated lifespan of the entire universe. So even if evolution had billions of years in which to work its magic (and it clearly does not in light of the new evidence that identifies many extinction events), natural mutations still could not generate the range of morphological changes Darwinists demand. This means that Darwinists have not yet identified the real factor driving innovation in life forms and that Darwinism still lacks a credible mechanism.

The biggest problem with mutations is, of course, the well-known fact that mutations are almost always either irrelevant to any possible evolution or fatal to the creature that manifests them. Regarding the first point, the DNA code is highly redundant. Code patterns that contain errors are normally ignored in the transcription process in favor of patterns containing the same information without the error. This means that mutations are normally passed over in transcription. The second point is made by asking a simple question: what parent reading this book would leap with joy on being told his or her child is a "mutant"? Darwinists would have us believe that it is by mutations that life evolves from microbe to microbiologist. In the real world, mutations mean a deformed body, pain, dysfunction, and an early death. Keeping these facts in mind is important because they help explain why no one has ever observed a new species rising from an antecedent species. If mutations are, as Darwinists insist, the only way in which such a thing could ever happen, then it is clear that such a thing has not been observed for the simple reason that it will, in fact, never happen.

Irreducible Complexity

Neo-Darwinism is a wobbly stool set up on three legs: natural selection, no saltus (gradualism), and random mutations. So far we have seen that there is a contradiction between observed facts and both the Darwinian rejection of saltus (they happen) and its last-minute appeal to mutations (there isn't enough time). The problem of irreducible complexity challenges both Darwinism's reliance on random transcription errors (mutations) and its insistence that natural selection is the one and only driving force behind evolution, ideas that are fundamental to the integrity of the theory.

Dr. Michael Behe, who has written the definitive study regarding this issue, defines irreducible complexity in these terms: "An irreducibly complex system cannot be produced…by slight, successive modifications of a precursor system, because any precursor to an irreducibly complex system that is missing a part is by definition nonfunctional."[11] Behe cites bacterial flagellum as an example of an irreducibly complex system. Flagellum are propeller-like rotary structures that make it possible for bacteria to move around. The operation of these structures requires the interaction of fifty different proteins all working together. If even one of these proteins is absent, the system is completely nonfunctional. *If we rely on Darwinist assumptions* it is difficult to see how the structure could have evolved since the benefit that supposedly provided a survival advantage did not appear until all fifty proteins were in place.

Those words, *if we rely on Darwinist assumptions* are important. Some evolutionists like to claim that Behe is arguing that irreducibly complex systems cannot evolve naturally. That is not true. The problem created for Darwinism by irreducibly complex systems is one that Darwinism brings down upon itself. Darwinism insists that evolution depends *entirely* on the operation of natural selection and that natural selection *only* responds to incremental morphological changes generated by random mutations. The problem is that *under those specified conditions* irreducibly complex systems cannot evolve. The reason is that as long as the complex system was only partially represented (which it would have to be for much of its development if it was generated entirely by the slow accumulation of *random* mutations) the mechanism under development would confer no behavioral advantage that natural selection could favor. It necessarily follows that if nature exhibits irreducibly complex mechanisms (and it certainly does), some other process beside natural selection and some other factor beside random mutations must be responsible for driving their development. That possibility robs Darwinism of its ability to claim that it offers a final and comprehensive picture of natural history and restores a good deal of mystery to the story of life.

This problem of irreducible complexity takes some interesting turns. Darwinists themselves are currently embroiled in a bitter dispute concerning what some call "nascent evolution." Many creatures, as part of their growth cycle, exhibit a larval stage that is quite different from their adult form. Sea squirts, for example, exist as fully formed tadpole-type creatures swimming in the sea before they settle down and become a potato shaped creature rooted to the ocean floor. These larval forms are complete beings with eyes, fins, and a digestive system. They lack only one thing: they have no organs of reproduction. How can the complex systems that sustain such creatures be credited to evolution if the creature itself *never* produces any young who could preserve the occasional segment of mutated DNA and pass it on to descendants under the blind gaze of natural selection?

Darwinists have made a few attempts to solve the problems posed by irreducible complexity but it is surprising how little they have actually written in this area. A few writers have speculated that some sort of "mutation storm" could fire an entirely new complex mechanism into immediate operation. Unfortunately, such an event would constitute an enormous saltus. Just think of the millions of bits of DNA information that would simultaneously have to be appear in perfect coordination within thousands of cells all without error and all at the same time! In terms of bits of information involved, such an event, which of course like everything else in Darwinism is a matter of pure conjecture—nothing like it ever having been observed in nature or the laboratory—would constitute an occurrence much less probable than any miracle story found in any religion. But that isn't the biggest problem Darwinists face. Just one credible example of such a mutation storm would permanently shift research away from the study of natural selection to the identification of the cause of that mysterious "storm." That kind of research would necessarily call into question the explanatory sufficiency of Darwin's little three-legged stool.

Some Darwinists have proposed a process they call "teleological evolution" as a way to resolve these problems. They argue that some intentionality built into nature pushes evolution through the unrewarding middle phases of the development of irreducibly complex mechanisms toward the final emergence of a new organ or behavior. Unfortunately, if such a claim could be proven, Darwinism would actually *become* creationism and Darwinists, who must then be banned from the public schools, would have to drag their "tree of life" charts into local Sunday schools. Teleological evolution eliminates the randomness of the system. It challenges Darwinism's core conviction that evolution is an *undirected natural process* and replaces it with the claim that a mysterious force is intentionally guiding the flow of natural history. In the view of a true Darwinist, such a proposal must be rejected as rank heresy. For many Darwinists, as we will see in a moment, the most appealing quality of the theory is the blessed assurance it offers that there is no God who can demand our obedience or judge our conduct. Teleological evolution destroys the foundation of that hope.

Confirmed, Plausible, or Busted?

Can we say today that Darwinism is falsified? Cable television airs a show that tests the validity of urban myths. To each myth they give a score: Confirmed, Plausible, or Busted. We can apply this approach to the three tests Darwin established for the falsification of his speculations.

1. *Fossil evidence must display continuous change with no stable forms.* The reality is that fossil evidence clearly demonstrates long periods of stability broken by sudden eruptions in which new biotic communities explode into life. The pattern of natural history is clear and it does not fit Darwinian predictions at all. This claim is busted.
2. Darwin insisted that *nature never makes leaps*. He did so because any leap would redirect attention from natural selection, which he claimed was the only agency driving natural history, to whatever force is causing the leaps. The possibility that unknown factors might play a role in directing natural history would deprive Darwinism of its claim to offer a unifying theory for biology. This claim was a nonstarter from the beginning if only because of the Cambrian explosion and its decisive effect on the history of life. Darwin himself admitted that. This claim is busted, hands down.

3. Darwin admitted that for his theory to work *a mechanism must be found* by which acquired characteristics could be passed down to progeny. Neo-Darwinists believed they had found such a mechanism in the fact that there are mutations or natural transcription errors in the transfer of DNA.

At first blush the neo-Darwinist proposal might seem plausible. It is clear that transcription errors do occur. It is also clear that such mutations can change the morphological characteristics of a creature. But we now know that such errors occur at a rate that is way too slow to power evolution. Moreover, mutations are virtually guaranteed to be destructive if they present any morphological expression at all. Finally, no credible proposal *that is consistent with Darwinian assumptions* has been offered to overcome the difficulties posed by structures expressing irreducible complexity. Those conjectures that have been suggested do as much harm to the fundamental structure of Darwinian thought as would be done by adoption of some form of creationism. It is, of course, always worth keeping the fact in mind that no case of a new species being generated by the processes described by Darwinians has ever been observed in nature or laboratory. The neo-Darwinist appeal to mutations therefore fails to save Darwinism, and the speculation as a whole is busted.

Some years ago, the miserable state of Darwinism was summed up by Colin Patterson, senior paleontologist at the British Museum of Natural History. At the beginning of a lecture, he offered this remarkable statement:

> It's true that for the last eighteen months or so, I've been kicking around non-evolutionary or even anti-evolutionary ideas. Now, one of the reasons I started taking this anti-evolutionary view, well, let's call it non-evolutionary, was last year I had a sudden realization. For over twenty years I had thought that I was working on evolution in some way. One morning I woke up, and something had happened in the night, and it struck me that I had been working on this stuff for twenty years, and there was not

> one thing I knew about it. That was quite a shock, to learn that one can be so misled for so long. So either there is something wrong with me, or there was something wrong with evolutionary theory. Naturally I know there's nothing wrong with me. So for the last few weeks, I've tried putting a simple question to various people and groups of people. The question is this: "Can you tell me anything you know about evolution, any one thing, any one thing that you think is true?"...The absence of an answer seems to suggest that it is true, evolution does not convey any knowledge, or if so, I haven't yet heard it. Now I think many people in this room would acknowledge that during the last few years, if you had thought about it at all, you've experienced a shift from evolution as knowledge to evolution as faith. I know that's true of me, and I think it's true of a good many of you in here. So that's my first theme. That evolution and creationism seem to be showing remarkable parallels. They are increasingly hard to tell apart. And the second theme is that evolution not only conveys no knowledge, but seems somehow to convey anti-knowledge, apparent knowledge which is actually harmful to systematics.[12]

In a later chapter, I will explain exactly why any theory of evolution will always have more holes than Swiss cheese. The problem has to do with a fundamental inconsistency between the premises underlying evolution and certain established laws of physics and with the fact that when you fight against physics, physics wins. My purpose in this chapter has been what I said it was at the beginning: to offer an exercise in unlearning. I wanted to establish that evolution is nothing like the 800-pound gorilla of manly muscular scientific fact that many people assume it is. I have cited only a handful of the many problems hounding evolution today (and have left out the most complicated issues). The fact is that the whole idea now teeters on wobbly legs. Colin Patterson had it exactly right; when it comes to Darwinism, it really is impossible to identify even one thing that is true anymore.

Sadly, many Darwinists have sought to bypass the painstaking search for "any one thing" that is true by saying many things that are not true. In his book *The Icons of Evolution*, David Wells recounts the long history of frauds and hoaxes perpetuated by Darwinists.

Wells outlines a history that stretches from Piltdown Man to the recent Piltdown Bird. He tells the story of Ernest Haeckel, a Basil researcher whose woodcuts supposedly proved that fetal development reiterates the evolutionary history of various adult creatures. Darwin himself described Haeckel's embryological argument as the "weightiest argument of all" in support of his theory.[13] To this day, textbooks still sometimes present Haeckel's claim as a proof of evolution even though his argument and his woodcuts were determined to be transparently fraudulent by the faculty of his own university. (Haeckel's phony drawings were, for example, reproduced in Douglas Futuyma's college textbook, *Evolutionary Biology*, as late as the 1998 edition.)[14]

Some of these deceptions are by now widely known, but a far more serious distortion passes almost unnoticed because it is so ubiquitous. If you examine any evolutionary "Tree of Life," you will see vertical lines tracing proposed lines of descent within a phylum or order. These lines are, of course, entirely conjectural insofar as the claim goes that the creatures portrayed stand in a historic relationship one to the other. The fact that two objects look like each other does not establish that one is related to the other by descent. But those vertical lines do generally show real fossils. These same Trees of Life also feature horizontal lines that connect the larger taxonomic units. What most people do not know is that all the lines connecting the phyla are completely hypothetical. The fossils represented on those lines are a product of something called *claddistics*.

Claddistics is the product of a completely unscientific faith in evolution. Evolutionists are absolutely convinced that their theory is true. When fossil evidence is lacking for a particular evolutionary claim, they feel justified in supplying said fossil to charts and diagrams in order to demonstrate a certain line of development. They reason that the fossils can be supplied because "the theory is true" so the fossils must be out there somewhere waiting to be found. Then these same researchers turn around and offer the pictures of the vaporware fossils they just invented as evidence that supports their theory. But is it really scientifically possible to do that? I, for one, hope so because if it is I can use the pictures at the back of my Bible to prove the resurrection of Jesus and all sorts of other interesting things as well. We could break out our bestiaries and demonstrate that unicorns really did exist. I wonder, do the guys with fake UFO pictures know about this?

The problem with all this is that scientists aren't supposed to fake data. Claddistic fossils exist only in imaginations heated by evolutionists' longings. The dark secret of evolutionary speculation is that no fossils have ever been found that connect the major phyla. Remember that the basic morphological types emerged all at once and out of nowhere in the Cambrian explosion. All the handy links connecting the phyla in biology books are a product of claddistics. They are generated by conjecture; they have never been observed in rock or soil and cannot be found in any museum. Once these conjectural lines and the imaginary fossils drawn along their planes are removed, what is left looks nothing like the proverbial Darwinist tree of descent working its way back to a common ancestor. Take away the phony art and what you see on the chart is a reflection of the strange pattern of mysterious appearances and sudden disappearances that characterizes the actual history of nature, a pattern that is not Darwinian in any sense.

When it comes to evolution, the driving question is why do evolutionists continue to cling to their speculation in spite of its failure as a scientific proposal? Why do these people continue to defend their conjectures to the point where many of them are willing to either manufacture evidence or embrace without examination any proffered evidence that seems to favor their position—evidence that often turns out to be fraudulent?

Scientific Evangelism

The answer has to do with philosophy and not science. Evolutionists often seem to be motivated not so much by the demands of evidence as they are by a deep aversion to God that verges on outright hatred. Richard Dawkins, one of the most prominent advocates of Darwinism, made this clear when he defined the purpose of his seminal book, *The Blind Watchmaker*. He wrote, "The purpose of this book is to provide a nonsupernatural explanation for the existence of complex living organisms."[15] It is important to see that Dawkins did not propose to offer a truthful explanation for how complex organisms came to be. He only promised *a nonsupernatural* explanation. The question of truth was for Dawkins secondary to the goal of removing the Creator from His creation. When for a brief time it appeared that evolution had found in random mutations the missing mechanism that might actually make it work as a scientific idea, evolutionist Jacques Monod expressed his joy in a way that demonstrated the real issue at stake, at least for him. He crowed, "The mechanism of Darwinism is at last securely founded. And man has to understand that he is a mere accident."[16]

Many evolutionists' commitment to atheism trumps respect for evidence—be it scientific, historical, or spiritual. If God stood up in front of these people at a Mount Sinai epiphany and said, "I created all this," they would reject the revelation. Their rejection would not be a deed done in service to truth, it would be a matter of faithfulness to a personal belief system to which they had already given wholehearted assent. Their point of view was put forward a few years ago by one of evolution's chief spokespersons, Richard Lewontin, who wrote, "Everything is negotiable except the vital objective of keeping God out of objective reality. We cannot allow a Divine Foot in the door. ...To appeal to an omnipotent deity is to allow that at any moment the regularities of nature may be ruptured, that miracles may happen."[17]

Note that Dr. Lewontin is not speaking here about rejecting bad evidence or clearing away false assumptions. He is expressing his deeply personal nonnegotiable commitment to the repudiation of God. He is prepared to lay aside any evidence or even abandon the search for truth altogether to pursue the eviction of the Creator. For him, this is a cause that is more urgent than the pursuit of scientific knowledge.

He is far from alone in harboring such feelings. Francis Crick, codiscoverer of the double helix structure of DNA, states the first commandment of the evolutionist as he sees it: "Biologists must constantly keep in mind that what they see was not designed, but rather evolved."[18] Here Crick is actually challenging biologists to squelch any nasty tendency to pay attention to the evidence provided by nature. Richard Dawkins seconded his challenge, writing, "Biology is the study of complicated things that give the appearance of having been designed for a purpose."[19] No matter what nature looks like, biologists, in this view, must maintain an absolute commitment to rejecting any concept of design not because such a concept is impossible or unscientific but because design might suggest the possibility of a Creator. My point here is not to defend the idea of design, my aim is to point out that for these researchers their personal beliefs trump scientific accuracy in the quest for truth.

Richard Dawkins is one of evolution's most effective spokespersons. His writing is interesting and unusual because it is honest and straightforward. Dawkins makes it quite clear that, for him, advocacy of an evolutionary point of view was never a matter of scientific investigation but was always something done in pursuit of a higher calling (as it were). Consider this statement and ask if this is in any sense the voice of science: "I want to persuade the reader not just that the Darwinian worldview happens to be true but that it is the only known theory that could, in principle, solve the mystery of existence."[20]

In real science is any theory ever the *only* option that can completely solve the *mystery of existence*? Of course not. But a theory can be the only explanatory possibility consistent with an unspoken agenda or prior commitment to a particular worldview and that is precisely the case with evolution. Consider in that light a comment by Richard Dawkins that we cited earlier in this chapter, noting his evangelistic zeal:

> In the context of the fight against Creationism, gradualism is more or less synonymous with evolution itself. If you throw out gradualness, you throw out the very thing that makes evolution more probable than creation. Creation is a special case of saltation—the saltus is the large jump from nothing to fully formed modern life. *When you think about what Darwin was fighting against*, is it

> any wonder that he continually returned to the theme of slow, gradual, step-by-step change?[21] (emphasis added)

This is not the language of disinterested science. Richard Dawkins is obviously a man on a crusade. In the end he is able to articulate his real motivation in unambiguously theological terms. "Darwin," he writes, "made it possible for me to be an intellectually fulfilled atheist."[22] We can appreciate Dawkins' honesty because he writes things that other evolutionists merely think. But we do have to question whether any scientific theory exists for the purpose of sustaining Richard Dawkins' religious commitments, or those of anyone else no matter their spiritual leanings.

The thing that must concern us is this: *if all the evidence ever needed to falsify Darwinism was known and published today, do you think Richard Dawkins and his like would tell you about it?* People whose dedication to Darwinism is based on deeply personal philosophical commitments rather than the examination of evidence will defend their speculations with all the force commanded by their prestige and the prestige of the institutions they control. The people quoted here are clearly motivated by their determination to "fight against creationism" and by their desire to be "intellectually fulfilled atheists." People so determined to defend a point of view will say whatever they have to say to convince others to believe as they do.

The problem is that scientists are not supposed to be partisans in what amounts to a theological debate. Much of the public supposes these men to be objective researchers, not evangelists. When a preacher confesses his faith, everyone understands that it is a faith that is being preached and they deal with it in that way. But when an evolutionist, supposedly "a man of science," makes his claims, people are apt to take that as a presentation of fact. But what if the scientist's statement expressed a preference that reveals only what s/he wishes was true rather than a statement of what really is true? In such a case, the scientist is clearly betraying the scientific spirit, but more than that, s/he is misrepresenting himself or herself to the public. Such a person can actually become dangerous, a barrier to the pursuit of truth. Backed by foundations, grants, prominent journals, and the uninformed parroting of mass media, "scientists" of this type may, for a time, sustain their position, but they will do so at the expense of authentic research and informed debate in their own field.

A Call for Clear Thinking

Fortunately, Darwinism is beginning to crumble. In July 2000, the Pew Forum on Religion and Public Life conducted a poll exploring public attitudes toward evolution and creation. The poll revealed that 42 percent of the public thought life existed pretty much in its present form from the beginning, a creationist position. Another 18 percent thought that evolution had occurred but that it was guided by God, a position that I define as a form of creationism. Only 26 percent of the public accepted the notion that the history of life was shaped by natural selection, the evolutionary position.[23] In August 2006, *Science* magazine published a poll that indicated the percentage of the population that accepted the idea of evolution had declined in the previous year from 45 percent to 40 percent. It is interesting to note that the percentage that absolutely rejected an evolutionary position had also declined from 48 percent to 39 percent, while the percentage who found themselves no longer sure about their position jumped from 7 to 21 percent.[24] While polls never determine the truth of a scientific or religious idea, these numbers are suggestive. They seem to indicate that the public is ready to welcome a new approach to life sciences, an attitude that must encourage anyone interested in the search for truth. When evolutionists must try to maintain their domination over public education by securing a court order that prohibits under force of injunction the classroom presentation of any scientific data that challenge their conjectures, an event that happened recently in Dover, Delaware, it is clear that the end is near for that monopoly.

It is not just the public that is demanding better research and more realistic thinking in the area of life sciences. While some evolutionists would have the public believe that only a few religious fundamentalists question Darwinian claims, in reality many secular scientists are now raising their voices in criticism of Darwinism. Recently, more than six hundred prominent biologists signed a letter indicating that they are in various ways skeptical of Darwinism. The signatories include Stanley Salthe, textbook author; Lyle Jenson, fellow of the American Association for the Advancement of Science; Richard Sternberg, Smithsonian Institution evolutionary biologist and researcher at the National Institutes of Health's National Center for Biotechnology Information; Giuseppe Sermonti, editor of *Rivista di Biologia,* one of the oldest currently published journals of biology in the world; and embryologist Lev Beloussov of the Russian Academy of Natural Science.[25]

Even a few serious researchers within the evolution community itself have recognized the need for new thinking. The list is impressive and includes such luminaries as Stuart Kaufmann, Ilya Prigogine, Manfred Eigen, and even Francis Crick.[26] Prominent science writer Stephen Jay Gould shared his personal experience on being forced to deal with the kind of problems outlined in this chapter. In the article "Is a New and General Theory of Evolution Emerging?" that appeared in *Paleobiology* 6:1, Gould wrote that the Darwinian theory "had beguiled me with its unifying power when I was a graduate student in the mid 1960s but since then I have been watching it slowly unravel as a universal description of evolution.... [Today] that theory, as a general proposition, is effectively dead, despite its persistence as textbook orthodoxy."[27]

The eventual fate of Darwinism had been predicted from the beginning and predicted by one of Darwin's closest friends and strongest supporters. No one did more to advance the ideas of Charles Darwin than T.H. Huxley. But writing in 1896 Huxley saw danger ahead for Darwin's ideas:

> History warns us ... that it is the customary fate of new truths to begin as heresies and to end as superstitions; and as matters now stand, it is hardly rash to anticipate that, in another twenty years, the new generation, educated under the influences of the present day will be in danger of accepting the main doctrines of the Origin of Species with as little reflection, and it may be with

> as little justification, as so many of our contemporaries, twenty years ago, rejected them.[28]

In his book, *The Fifth Miracle*, scientist and philosopher Paul Davies examined the current state of biology. He noted the problems facing Darwinism but went beyond that to point out some facts that are not widely known. He reminds us that after a hundred years of experiments subjecting every sort of amino acid mixture to every imaginable kind of energy, scientists working within the Darwinist framework have not been able to synthesize a single protein molecule. We are not talking here about generating life or designing a cell. We are talking about synthesizing just one protein molecule. Think about that the next time some scientist is quoted in the newspaper confidently describing the origin of birds or outlining in detail the developmental history of whales. We must also remember that the evolution of one species into another *within the framework of mutation and natural selection proposed by Darwinists* has never been observed in either nature or the laboratory. We have observed oscillation of beak sizes in finches and alternation in dominant color traits among moths in England. But the emergence of hitherto unexpressed traits within an already existing population is not the Origin of Species.

Paul Davies summarized a study of recent trends in life sciences by suggesting that to move forward biology needs some radically new ideas. Then, in an article titled "Bit Before It?" he focused on one very promising new approach to life sciences research:

> Normally we think of the world as composed of simple, clod-like material particles, and information as a derived phenomenon attached to special, organized states of matter. But maybe it is the other way around: Perhaps the universe is really a frolic of primal information, and material objects a complex, secondary manifestation.[29]

That observation launches us toward a better way of thinking about natural history. In this book we have been moving toward an ability to read a radical scripture, the first chapter of Genesis, for exactly what it says. To get to that point we had to clear away an oddly persistent nineteenth-century enthusiasm known as Darwinism. Still moving toward our goal we will now follow Paul Davies' prescription and turn to a research field that is at once as current as twenty-first-century computer science and yet at the same time an approach to knowledge that draws us back to the dawn of science itself, the field of study known as Information Theory.

Chapter 3

Pythagoras and the Age of Information

Sometimes a single moment can provide more enlightenment than a dozen books. Such a moment happened a few years ago when Richard Dawkins was being interviewed for a television production. As that interview unfolded, Dawkins was asked if he could cite one example, just one, in which natural selection had generated new information. In response to that question, Dawkins sat silent for eleven seconds, an eternity in television terms. After this startling pause, Dawkins stammered out a response that had nothing to do with the question. That widely reported and videotaped episode sets us on the trail that leads into the book of Genesis. It does so because it illustrates the difference between the nineteenth-century approach to science upon which Darwinian speculations rest and twenty-first-century science.

Biology in the nineteenth century was not very complicated. The biologist was asked to observe the physical appearance of a thing, for example a living cell, and then describe what he saw. The internal operation of that object was mostly left to the imagination. Generally, scientific research 150 years ago began with observations and measurements focused on the superficial appearance of a thing, which then led to speculations about what might be going on beneath the surface.

The twenty-first-century approach is different because our tools have improved. We are no longer limited to describing outward appearances. We are now able to study the internal operations of everything from a continent to an atom. As we have looked more deeply into the heart of the material world, we have discovered that the essence of everything is not defined by the physical form of structures and systems but consists in the exchange of information. A good example is afforded by life itself. In biology, this fresh approach to the study of living organisms began with the discovery and unraveling of the operations of DNA.

If you are willing to have your breath taken away, it is worthwhile finding a textbook and reading a detailed description of the processes by which DNA copies itself, edits its data, and so synthesizes a protein molecule. It is even more amazing to read how DNA unwinds itself, cuts itself, and repacks itself during mitosis, all at speeds of eight thousand revolutions per minute and all of it accomplished without twisting or tearing. But for the purpose of our journey into Genesis, the thing we need to notice is that all these complex operations are directed to a single task. That single task is the generation and communication of *information*.

What, after all, is deoxyribonucleic acid? DNA is a digital code. It is information written not with ink but with four nucleic acids that draw from an alphabet of twenty amino acids to assemble more than 100,000 different proteins that, in turn, provide the physical foundation of life. This information system has several characteristics. It is able to monitor internal communications and instantly correct transcription errors. It includes a system of punctuation. It even contains what information theory calls *conflict resolution logic*.

The basic insight of twenty-first-century biology can be expressed as follows: if we want to understand life we have to understand the operations of DNA. To do that we have to understand how codes work because DNA and the life processes it governs are entirely a matter of coding and decoding information.

In order to communicate information, all codes require two elements. One element is physical. There must be some sort of template on which the code is written and a medium that is applied to that template. This is the paper and ink of coded communication. In the case of DNA, these physical properties are supplied by amino acids, nucleic acids, sugars, phosphates, and proteins.

The other element needed for coded communication is conceptual. There must be a message that is expressed through the code. It is, after all, for the purpose of communicating a message that any code exists. In a system involving coding and decoding, the message is far and away the most important thing.

The Origin of Code

It is at this point that we arrive at the distinction between Darwinian speculation and twenty-first-century science. Neo-Darwinism is focused 100 percent on material substances, the paper and ink of life. It can't even acknowledge the existence of nonmaterial factors like information. Because of its assumption that only material things exist, Darwinism must insist to the bitter end that ink and paper write life's book. It is never explained how they might actually do that since there don't seem to be many books authored by their own paper and ink. Darwinism must simply insist that this is how it happened because Darwinists, believing as they do that physical reality is the only reality, have nowhere else to turn for the source of the information that organizes life.

In actual practice this means that under Darwinian assumptions we must presume that either DNA spontaneously generated itself or living organisms spontaneously generated themselves without DNA. This is a necessary conclusion because living organisms are, as far as we know, the only entities in nature that generate DNA. But the idea is also problematic because, again as far as we know, living organisms depend on DNA for their existence. We are, then, left with the conclusion that life must either have spontaneously generated itself before the DNA that makes life possible existed or DNA must, somehow, have organized itself before the living beings that produce it appeared. Neither proposition works because the concept of spontaneous self-generation on which this whole line of speculation depends has not been an acceptable scientific principle since the time of Louis Pasteur.

Post-Darwinian research approaches the matter of the origin of DNA differently. It understands that in the processes of life the crucial factor is the information that organizes how the ink is applied and gives content to what appears on the paper. In other words, we now see that the arrangement of nucleotides on a chain of DNA only has meaning to the degree that it communicates *a message* that results in the formation of a specific kind of protein. The mystery of life, then, hides not in the physical arrangement of the chains of nucleotides (the ink and paper) but in the source of the message that their arrangement communicates. DNA is a code and a code exists only to communicate a message. The challenge facing researchers today can be stated as follows: if we can find the source of the code we will have found the structuring agent that generated DNA and in identifying that agent we will have discovered the real author of life.

Physicist Werner Gitt, director of the German Federal Institute of Physics and Technology, described the challenge this way: "He who can give an account of the origin of information holds in his hands the key to interpret this world."[1] This focus on the origin and content of information has changed the way scientists go about their work. The reason for the change is that while matter can receive information, it cannot create information. Paper and ink don't write books because matter can't generate information. German physicist J. Peil underscored this point when he wrote, "Information is neither physical nor a chemical principle like energy and matter, even though the latter are required as carriers." [2]

This repositioning of research has created a new kind of natural science in which the description of physical structures and systems gives way to analysis of the information that structures the operation of those systems. In this new science, microscopes are replaced by computers. The ability to generate pictures and diagrams of material things, for so long crucial in biology, is replaced by the need for skill in the complex mathematical formulas of information theory.

This turn toward analysis based on information theory is part of the revolution tearing at the dying roots of Darwinism. Information theory changes the Darwinian question, which is always some version of "Why does an elephant have a trunk?" into a new question, "Where does the information on which any physical structure or system is based come from?" It is at this point that the story about Richard Dawkins' embarrassment becomes significant. Darwinism has no answer to the question about the generation of information.

Some years ago, the International Society for the Study of the Origin of Life (ISSOL) held its annual meeting in Mainz. Reporting on that event, physicist Klaus Doss wrote, "A further puzzle remains, namely the question of the origin of biological information, i.e., the information residing in our genes today." Doss then referred to Louis Pasteur's address to Sorbonne University in 1864 in which Pasteur predicted that as a result of his work the theory of the spontaneous self-generation of living cells had suffered a blow from which it would never recover. Professor Doss makes the same prediction for the effect of the Mainz meeting on the theory of evolution. "The Mainz report may have an equally important historical impact, because for the first time it has now been determined unequivocally by a large number of scientists that all evolutionary theses that living systems developed from polynucleotides which originated spontaneously, are devoid of any empirical base."[3] This conclusion means that the secrets of life will be discovered not through a microscope or at a fossil dig but by exploring how *ideas* are formed and communicated, topics that are the focus of information theory.

Information theory is a field full of specialized jargon and complicated rules. Fortunately, there is a way to learn how an information-centered approach to science works and at the same time continue to move steadily toward our encounter with Genesis. We can do that by journeying back through time to a philosophical community that viewed all problems in physics and biology as a matter of ciphering and deciphering digital codes, a community that focused on the study of pure information 2,500 years before the term *information theory* was coined.

A Numbers Game

The school I speak of was associated with a man named Pythagoras. Pythagoras was born on the island of Samos around 580 BCE. In 530 BCE he settled in Croton, a town in southern Italy. The years between those dates must have been exciting. It is reported that Pythagoras studied geometry and astronomy with the Ionian scientist Anaximander. He also studied symbolic mysticism with priests in the temples of ancient Egypt. Interestingly, it is widely reported that he studied dream interpretation and other sacred subjects with Hebrew sages. He journeyed among Arabs and Chaldeans and, finally, met the prophet and teacher Zoroaster.

At the end of his travels he founded a community in Italy. His followers called themselves Homakooi, "Those who come together to listen." The Homakooi were an exclusive group. Like the members of the early Christian church they held their property in common. Everything they heard in their teaching sessions was deemed a closely held secret. No one was permitted into their meetings until they had completed a rigorous period of training, again in a manner similar to the early church where the full mysteries of the faith were made known to initiates only after months of participation at the edges of the community. It appears that early Pythagoreans observed a vegetarian diet (as did many Christians). At a later date, the sect was allowed to eat meat but only from animals offered in sacrifice to the gods. Pythagorean women were treated as equals and sex between married couples was deemed a specifically religious obligation. Some Pythagorean rubrics seem to have been quite strange. Pythagoreans could help someone load but not unload items, they could not eat beans, and they could not break bread. The problem with bread had to do with the fact that it was a symbol of friendship that should not be broken. There is no explanation given for the problem with beans.

What is relevant here about this strange group of people is that they taught a vision of the universe in which they held that all things are, at their root, made entirely of number and proportion. The first person to set the secret Pythagorean teachings to paper, Philolaus, put it like this: "All things which are known have number; for nothing can be known or understood without number."[4] By this the Pythagoreans did not mean simply that everything can be counted. They meant that everything was actually *generated* by powers and qualities inherent in numbers. Aristotle, one of many ancients who studied Pythagorean thought, described their position. He said, "Pythagoreans construct the whole heavens out of numbers."[5] For Pythagoreans the universe began with a blueprint that had no material content whatsoever but consisted entirely of numbers. "The first principles of these (i.e., numbers) were the first principles of all things. In numbers rather than in fire and earth and water they thought they saw many likenesses to things that are and that are coming to be."[6]

One expert on Pythagorean thought, Charles H. Kahn, describes the process of creation, as the Homakooi understood it, this way:

> Cosmogony begins as the numbers are generated when the Unlimited is drawn in (or "breathed in") by the limiting principle. Thus the cosmos arises from the One by breathing, like a newborn animal. The heavens take shape as breath (pneuma) and void are drawn in from the Unlimited, as a separation and differentiation of things within the sphere. *"And this happens first in the numbers"* [quoting Philolaus; emphasis added].[7]

The language of Pythagorean thought is strange but the basic concept is simple. In Pythagorean philosophy everything that can be seen in the material world exists first in the form of a mathematical blueprint. For Pythagoreans, that blueprint constitutes reality as opposed to materiality, which is deemed to be only a shadow. This blueprint initially takes the form of a sequence of numbers. For example, the concepts of "limiting" and "unlimiting" principles mentioned in the above quotation referred to odd and even numbers, respectively.

Simply offered as an illustration of how a Pythagorean would think, consider the following example. The number 5 was especially sacred in the Pythagorean system. This was so because the relationship $3^2 + 4^2 = 5^2$ defined the regularity of right triangles. In this relationship, the number 5 was known to Pythagoreans as "marriage" because it joined the numbers 3 and 4 and so linked the realm of the limiting (odd numbers) with the realm of the unlimiting (even numbers). For Pythagoreans, the triangle was a fundamental structure that gave the cosmos its physical form and defined the operations of life because it expressed a relationship between fundamental categories of numbers. The important thing to notice is that the triangle and the material cosmos to which it mediates existence are, for Pythagoreans, both rooted in an ultimate reality made of nothing but pure numbers, a nonmaterial approach to cosmology that we will soon encounter in Genesis.

Cosmology as Code

Pythagorean philosophy is a vision of the universe in which material reality only plays a secondary role. This is a cosmology that sees ultimate reality as a kind of software that exists entirely in the form of a digital code. As to the question of where information (the code) comes from, Pythagoreans had a ready answer. The source of the information was God, a God they described as a Trinity:

> The first god is simple and remains in himself since he holds converse entirely with himself he is in no way divisible. The god who is second and third, however, is one. In associating with matter which is a dyad, it provides unity to matter while being at the same time separated (or split, *schizetai*) by matter, since the latter is in flux and has an appetitive character.[8]

Pythagoreans, in other words, speak of a form of God who is beyond space and time and two forms of God who are either incarnate or participate in the material world and that are in some measure humanized and limited because of that participation.

Pythagoreans went into quite a bit of detail in describing the process by which numbers were translated into physical "appearances" (the Greek word referring to the appearance of something is best translated as "seemings" in such a way as to suggest that everything that appears to be a material object is actually rooted in the subjective experience of an observer). In Pythagorean thought, the physical world is made up of four or in some cases five elements, the classic elements of Greek physics: earth, air, fire, and water (and sometimes ether). The first four of these elements are, in turn, considered to be the material form of geometrical abstractions. According to the Pythagorean philosopher Hermes, each of these elements and their geometrical blueprint is constructed entirely out of triangles, an idea we will also find expressed much earlier in the text of Genesis.

> Fire is composed of twenty-four right-angled triangles, surrounded by four equilaterals. And each equilateral consists of six right-angled triangles, which they compare to the pyramid. Air is composed of forty-eight triangles, surrounded by eight equilaterals. And it is compared to the octahedron, which is surrounded by eight equilateral triangles, each of which is separated into six right-angled triangles so as to become forty-eight in all. And water is composed of one-hundred-and-twenty triangles, surrounded by twenty equilaterals, and it is compared to the icosahedron, which is composed of one-hundred-and-twenty equilateral triangles. ...And earth is composed of forty-eight triangles, and is surrounded by six equilateral pentagons, and it is like a cube. For the cube is surrounded by six tetragons, each of which is separated into eight triangles, so that they become in all forty-eight.[9]

These ideas may sound complicated but the main idea behind them is very simple. Pythagoreans demonstrate a way of thinking that consistently defines reality in terms of information, specifically *digital* information presented in the form of geometrical figures. This is a cosmology that does not pay *any* attention to the physical world of material existence. In the Pythagorean system, everything is numbers and numbers are everything. No one summarized the Pythagorean approach to reality better than Galileo did in describing his own view of how the universe works:

> Philosophy is written in that great book which ever lies before our eyes—I mean the universe—but we cannot understand it if we do not first learn the language and grasp the symbols in which it is written. This book is written in the mathematical language,

> and the symbols are triangles, circles, and other geometrical figures, without whose help it is impossible to comprehend a single word of it; without which one wanders in vain through a dark labyrinth.[10]

The fifteenth-century mathematician Nicolas Cusa summarized the thinking of Pythagoreans even more precisely when he wrote, "Number is the first model of things in the mind of the Creator."[11]

Two last points regarding the Pythagorean method of constructing a cosmology will set us on a course leading directly into Genesis. First, for the Pythagoreans, the number seven held extraordinary significance. The Jewish mystic Philo, whose work was influenced by the Pythagoreans, expressed their point of view when he wrote, "I do not know if anyone can adequately sing the virtues of the number seven, for they are greater than any rational account."[12] This attention to the first seven numbers was connected to a correspondence intuited by the Homakooi between the structure of the cosmos and the mathematical ratios underlying the seven tones of a musical scale, an intuition preserved today in the famous notion of "the music of the spheres."

There is one final point. Pythagoreans created a sophisticated cosmology in which the operations of the universe were determined by the ciphering and deciphering of a digital code. They created a physics and a biology built upon the study of nonmaterial mathematical principles rather than the analysis of material objects. The question must be asked, where did this information-centered cosmology come from? Can we trace the roots of this digitally based approach to natural history?

Students of Pythagoras have consistently pointed to one source as particularly important in shaping his thinking. That source is the book of Genesis. Ancient writers had long made it known that Pythagoras had studied with Hebrew sages. The historian Malchus reported that "[Pythagoras] spent time with Zabratus, a Jew, who purified and cleansed him from the sins of his former life."[13] Other ancient scholars who insisted that Pythagoras relied on Hebrew wisdom in shaping his teachings included Hermippus, Origin, Aristobulus, and Porphyry.[14] When we remember that Plato, the father of western philosophy, has been called a disciple of Pythagoras, it is significant that the neo-Platonist, Numenius, posed the question, "Who is Plato but Moses speaking Greek?"[15] But it was the great astronomer, Johannes Kepler, who saw most clearly the connection between his Lutheran faith and the thoughts of the Pythagorean school via a common root in the words of Genesis. Commenting on Plato's work, the *Timaeus*, the most purely Pythagorean of all Plato's texts, Kepler wrote, "This work is, beyond all doubt, a commentary on the first chapter of Genesis, otherwise the first book of Moses, transforming it into Pythagorean philosophy, as will easily be apparent to an attentive reader who compares it with Moses' own words."[16]

Thanks to the ancient Pythagoreans, we now have some sense of how an information centered approach to cosmology can work. With that experience in mind, we will, in the next chapter, turn to the original information-centered cosmology, found in Genesis 1, and explore the message of this truly radical scripture. We are at last prepared to read Moses' own words.

Chapter 4

Quantum Genesis: Reading the Language of Life

We have readied ourselves to enter the thought world of Genesis by exploring the idea that radical scriptures have valuable things to say provided that we learn how to read them accurately. We have seen that the best way to approach radical Bible texts is by means of strictly literal interpretation. If we fail to read a challenging text literally we will not likely receive from that text a message that can renew our minds or even attract our interest. We are, of course, always free to accept or reject whatever the Bible is teaching. We are not, however, free to tame its message by interpreting the text in such a way as to remove anything offensive or strange, because that sort of exercise leaves us with nothing more than the opinions and ideas we hoped to transcend.

The cosmology of Genesis rests on two big ideas. The foundational claim of Genesis is summed up in the first sentence: "In the beginning God created the heavens and the earth" (Gen. 1:1). This sentence tells us some things but it leaves many other questions unanswered. It tells that the universe was created and by whom but it does not tell us how it was made. For that reason it does not tell us much about what kind of universe we inhabit.

The second big idea in Genesis is its claim that the universe was created in six days. This claim begins to add detail to the cosmology offered by Genesis. It tells us that the universe was not a product of natural development, random accident, or evolution. The universe was the result of divine initiative exercised within the framework of a single work week. The six days of Genesis are not geological epochs. There is no textual evidence to support that kind of speculation. The six days of creation are six twenty-four hour days as measured on earth.

Evidence that this is how Genesis must be read is found throughout the Bible. One key passage is found in Exodus 20, where God says:

> Remember the Sabbath day, to keep it holy. Six days you shall labor and do all your work, but the seventh day is the Sabbath of the Lord your God. In it you shall do no work; you, nor your son, nor your daughter, nor your male servant, nor your female servant, nor your cattle, nor your stranger who is within your gates. *For in six days the Lord made the heavens and the earth, the sea, and all that is in them, and rested the seventh day*. Therefore the Lord blessed the Sabbath day and hallowed it.
>
> —Ex. 20:8–11, emphasis added

The Exodus passage clearly speaks of six normal earth days when it enjoins six days of work and a day of commemoration every seventh day. It makes no distinction between that kind of week and the seven-day week described in Genesis 1. It follows that the days of creation remembered every seventh day constituted an ordinary earth week.

Even stronger proof that the Bible speaks of a literal seven-day creation that included man's entry on the stage of history comes to us in the New Testament. St. Paul summarizes the story of the redemption of the human race in straightforward terms: "Therefore, just as through one man sin entered the world, and death through sin, and thus death spread to all men…much more the grace of God and the gift by the grace of the one Man, Jesus Christ, abounded to many" (Rom. 5:12, 15).

This is a very important text for understanding the message of Genesis. The Bible tells us that death entered the world only through the sin of Adam. It also explains Jesus' sacrifice on the cross in light of that claim. If life evolved through eons of development and extinction, then death must be considered an ordinary part of life and not a result of human sin. But if this is so there was no need for the cross, and the death of Jesus Christ accomplished nothing. All of this would imply that Jesus Christ, who foolishly gave up His life for no reason, could not have been the creator of a universe that, it turns out, He did not understand, in which case the gospel message falls apart.

Any theory that proposes that the physical universe developed over years full of catastrophe and extinction prior to the appearance of man is incompatible with the Christian faith. The problem does not have to do with a conflict between such claims and the Christian doctrine of creation. The issue is that a cosmology embracing such a period of physical development cuts the heart out of the Christian doctrine of salvation by introducing the reality of death prior to the fall of Adam. The biblical claim that God created the universe and did so in six days culminating in the sudden appearance of man who immediately and profoundly changed the structure of the universe is nonnegotiable for people who call themselves a Christian.

Does this then mean that every Christian is bound to accept the arguments of the group known as "young earth creationists," who tell us that every hill was raised and every animal was formed, covered with fur, and sent scurrying on its way a mere five thousand years ago? The answer to that question is no, Genesis does not require or even encourage us to accept arguments of that kind. But to understand what Genesis does want to teach us we need to pause and make some distinctions.

Material Reality

As we delve into the text of Genesis, the first thing we need to do is to take a closer look at a way of thinking called *philosophical materialism*. We need to master the ideas of this nineteenth-century school of thought because if we want to gain anything from Genesis we need to be able to understand materialism well enough to be able to think in exactly the opposite way. Many people assume that they have a problem accepting the Genesis account of creation because the Genesis account is naive or simplistic compared to the studied sophistication of science. This is not true. The real problem these readers are struggling with is philosophical. Genesis does not share the premises accepted by those who adhere to a materialistic paradigm. Materialism and the text of Genesis offer utterly incompatible pictures of reality.

Materialism is the belief that physical phenomenon are the *only* reality that ever was, is, or will be. Materialism assumes that every real event takes place in the material world and that every event always has a physical cause resulting in a physical effect. The world experienced by adherents to the materialist philosophy is a closed system in which physical events produce other physical events in a tight circle that admits no influence from external or nonmaterial factors because it is presumed from the beginning that such factors do not exist. Every form of materialism is by definition atheistic.

It used to be that most scientists were materialists, especially during the nineteenth century when the theory of evolution was proposed. As we will see shortly, this is no longer the case. Materialist philosophy actually poses serious problems for many people trying to do scientific work today. Because materialism operates within a tightly closed circle in which every cause and every effect is always physical, the philosophy runs into difficulties when faced with certain questions, such as those related to origins. Since materialism cannot appeal to any influences outside its little circle of physical causes and physical effects, when it deals with questions of origin it always ends up proposing some form of spontaneous self-generation. We have shown how materialist biology must assume that either DNA or living organisms generated themselves spontaneously. In the same way, a materialist must assume that the universe as a whole somehow created itself, an event that critics of materialism have derisively labeled the "Big Bang." I pointed out that the problem with all this is that we have known since the time of Louis Pasteur that spontaneous generation is simply not a scientifically valid idea whether applied to living cells or nascent universes.

In spite of difficulties like these, many people continue to insist that materialism is the only philosophical position appropriate for scientific work. Some go so far as to equate philosophical materialism with science itself. This, of course, is absurd and we will examine a very important contemporary nonmaterialist form of physics in a moment. But the claim made by some people that scientific work requires a researcher to accept the premises of philosophical materialism does pose an interesting challenge to science itself. *Is science the search for truth or is science merely an effort to concoct some sort of materialistic explanation for phenomena, regardless of what the truth might be?* If science really is nothing more than an effort to cobble together an explanation for phenomena no matter how unlikely that explanation may be just so long as it does not violate any of the parameters of philosophical materialism, then science is not part of the search for truth. That kind of science would be nothing more than a parlor game played for the amusement of card-carrying materialists. If science is really the search for truth it cannot restrict itself to explanations consistent with one philosophical standpoint, and certainly not the narrowest and least useful of all positions, the old-line nineteenth-century philosophy of materialism.

This is a good point in the development of the argument made by this book to take a look at the school of thought known as creationism. *Creationism* is a word that is thrown around these days without much effort to be precise about its meaning. Evolutionists try to use the word to refer to any set of ideas that challenges their materialistic philosophical premises. This is a demagogic misapplication of the term. In the real world, creationism refers to a very specific idea, namely, the notion that the *material* universe was created in a six-day period and that its origin took place only a few thousand years ago. Note the emphasis on the word *material*.

Let me begin by admitting that from a biblical point of view, some of the contentions of "young earth creationists" are correct. As I have shown, the Bible does teach that the universe was created over a literal six-day period. The Bible itself gives no direct indication as to the time when the work described in Genesis took place, but that is not where creationism goes wrong. Young earth creationism must also be credited with trying to deal with the problem posed by evolutionists' claim that death and destruction have been a natural part of life for billions of years. I spoke of this earlier. In the biblical worldview, sin, death, and their cosmic effects are all introduced into a deathless world by the fall of Adam and Eve. If this perspective is not affirmed, the sacrifice of Jesus Christ makes no sense. Young earth creationists are aware of this problem and try to address it. Those so-called "old earth creationists" and "Christian evolutionists" who welcome the notion that the *material* universe that we know today has been around for billions of years unwittingly turn Jesus' sacrifice into a meaningless gesture on the part of an ignorant man.

Created Reality

The problem with creationism is that it begins well but quickly abandons the premises that actually inform the biblical worldview. In attempting to expound a text (Genesis) that takes a decidedly nonmaterial approach to natural history, creationism interprets that text in terms that are as deeply rooted in the thin premises of philosophical materialism as anything written by Richard Dawkins. Creationism grabs a text that focuses on the encoding of information and twists it into a description of the nuts and bolts construction of physical objects. The two questions, where did information come from and where does matter come from, are, as we have seen, completely different topics. Creationists confuse these topics and as a result their research becomes a collection of wild speculations full of impossible notions and implausible conjectures that find no expression in either the text of the Bible or the evidence of nature.

Creationism, for example, offers the completely nutty idea that Genesis is speaking in materialistic terms when it tells us that grasses, herbs, and fruit trees were all flourishing on earth one full day (day three) before the sun began to shine (day four). Respect for the integrity of the biblical text should lead a person to suspect that at that point Genesis must be describing a process that is very different from anything that materialist philosophy would read into the text. Does anyone really believe that Jewish farmers, people intimately familiar with how plants grow, would have accepted the Genesis order of creation (plants first, sun later) as a description of physical events in a material world? Creationists would do well to remember what St. Augustine wrote in the fourth century:

> It very often happens that there is some question as to the Earth or the sky, or the other elements of the world … respecting which one who is not a Christian has knowledge derived from most certain reasoning or observation, and it is very disgraceful and mischievous and of all things carefully to be avoided that a Christian speaking of such matters as being according to the Christian scriptures, should be heard by an unbeliever talking such nonsense that the unbeliever perceiving him to be as wide of the mark as east is from west, can hardly restrain himself from laughing.[1]

Because of the publicity given to young earth creationists, many people assume that faithfulness to the message of Genesis requires that they abandon everything they know about the natural world and take sides in an imagined conflict between the Bible and science. This is not at all the case. The conflict people face is not between science and the Bible; the choice they are asked to make falls between materialist and nonmaterialist premises, either of which can provide a framework for serious scientific work. We also would do well to remember that the Bible is presenting the history of the universe from a perspective altogether different from any other approach to that history. Its account differs from other accounts because secular cosmologies always reflect a human perspective while the perspective from which the biblical record was written is self-consciously Divine. Genesis tells us what the Creator wants us to know about the internal operations of our world, how things got organized, and how the universe really works, information we could not otherwise gather by ourselves.

When we understand the uniqueness of Genesis and look at other approaches in its light, the main thing we notice is not the difference between evolutionary science and creation science but rather their remarkable similarities. Both of those enthusiasms depend heavily on fanciful speculations, windy explanations, and lots of fake fossils. Why are the two approaches so alike? The two schools of thought are similar because they share the same, basic (false) premise. They share an infatuation with philosophical materialism, the presumption that questions of origin must be resolved in grossly physical terms because material reality is the only form of reality that exists.

Evolution and creationism are, philosophically speaking, the same thing. They are both bastard children of the nineteenth century's romance with materialism. Their dispute is an argument inside the same dysfunctional family. It is a dispute that need not interest us in the twenty-first century because we have moved on. Materialism is no longer a dominant force in the science that has always mattered most, the discipline of physics. If you have a grasp of history you know what that means. It means that materialism is on its way out as a cultural convention. Popular paradigms always mirror contemporary physics with several years of lag time built in. Evolution is in trouble today not because of creationism or intelligent design but because that lag time has just about run out and the effects of twentieth-century nonmaterialist physics are starting to be felt throughout society. In the television series *Star Trek*, Montgomery Scott, chief engineer aboard the Starship Enterprise, summed up the problem facing all forms of materialist science very well when he spoke those famous words, "But I canna change the laws of physics, Captain!"

The real enemy of creationism is the same enemy that evolution will eventually have to face. That enemy is a body of emerging approaches to solving long-standing problems in physics that are not premised on materialist philosophy. Chief among these new approaches is quantum physics.

Observational Reality

Quantum physics is based on the discovery that a long-cherished model of the atom gave a false picture of the subatomic world. You may remember that model from high school science class charts that pictured the atom as something like a solar system with a sun in the middle (the nucleus) and various little balls called electrons flitting around it like planets. Quantum physics discovered that this picture was wrong. It turns out that all these little balls were only the visible expression of a deeper reality. Normally an atom does not actually display the solar system appearance shown on the old charts. An atom can best be thought of as a number of slots. Each slot is a preset level of energy known as a *quantum*. As levels of pure energy, quanta normally have no physical existence, but each quantum can temporarily take the form of a discrete particle under certain conditions that we will examine in a moment. When quantum physics describes these quanta as *preset* energy levels, it is telling us that energy does not glide between quanta like the rubber of a balloon gradually expanding. Energy jumps between the slots and only exists within the parameters of a certain slot. This jump is what is known as a *quantum leap*; a leap from one discrete energy level to another unrelated energy level.

The really shocking discovery made by quantum physicists was the realization that these quanta or preset energy slots are the normal state in which reality exists. In other words, the little things we call particles that serve as the foundation of the physical world normally exist as numbers on a mathematical scale. These scaled numbers, which represent the likelihood that a particle will appear at one particular energy level, are the permanent form in which all things exist, but since these probabilities are numbers, they have no physical form or material substance at all. Dr. Amit Goswami, professor of physics at the Institute of Theoretical Sciences at the University of Oregon, lyrically describes the normal state of an electron this way: "Between observations the electron exists as a possibility form, like a Platonic archetype, in the transcendent domain of potentia. (I dwell in possibility wrote the poet Emily

Dickinson. If the electron could talk, this is how it would likely describe itself.)"[2]

If reality consists in the abstract form of mathematical possibilities, where do particles like protons and electrons that provide the foundation for material reality come from? The Copenhagen Interpretation of quantum physics, the mainstream definition of the science, tells us that material particles only come into momentary existence in the presence of an observer who is actively taking measurements. Dr. Goswami explains the idea this way: "Can we assign the electron any manifest reality in time and space between observations? According to the Copenhagen Interpretation of quantum mechanics the answer is no."[3] This does not mean that the particles and the physical world they define are not real. It does mean that the particles that form material reality are a secondary phenomenon, a temporary and contingent artifact that results from the activity of an observer who is actively taking measurements.

Michio Kaku, Henry Semat Professor of Theoretical Physics at the Graduate Center of the City University of New York, reduces the technical foundation of quantum physics to three points:

> 1. All energy occurs in discrete packets, called quanta … .
> 2. Matter is represented by point particles, but the probability of finding the particle is given by a wave… .
> 3. Before an observation is made, an object exists in all possible states simultaneously. To determine which state the object is in, we have to make an observation, which "collapses" the wave function, and the object goes into a definite state. The act of observation destroys the wave function, and the object now assumes a definite reality.[4]

It becomes immediately apparent how important this discovery was when we remember that everything in the material world is made of subatomic particles like electrons. Science writer John Gribbon summarizes the shocking implication of the discovery announced by Max Born, Niels Bohr, and Erwin Schrodinger at Copenhagen in 1927 with a single sentence: "Nothing is real unless we look at it, and it ceases to be real as soon as we stop looking."[5] It is important to remember that the term *real* in this context refers only to reality in its material context. In quantum theory reality, is a thing much bigger than its physical expression in the material world?

When Albert Einstein was confronted with the conclusions of quantum physics, he reportedly cried out in disbelief, "Do you mean to tell me that the moon only exists when I look at it?" Einstein did not like the answer he received and you might not either, but this is the answer to Einstein's question as summarized by Dr. Goswami:

> To the extent that the moon is ultimately a quantum object (being composed entirely of quantum objects), we must say no—so says physicist David Mermin. Between observations, the moon also exists as a possibility form in transcendent potentia. Perhaps the most important, and the most insidious, assumption that we absorb in our childhoods is that of the material world of objects existing out there—independent of subjects, who are the observers. There is circumstantial evidence in favor of such an assumption. Whenever we look at the moon, for example, we find the moon where we expect it along its classically calculated trajectory. Naturally we project that the moon is always there in space/time, even when we are not looking. Quantum physics says no. When we are not looking, the moon's possibility wave spreads, albeit by a minuscule amount. When we look, the wave collapses instantly; thus the wave could not be in space/time. It makes more sense to adopt an idealist metaphysic assumption: *There is no object in space/time without a conscious subject looking at it.*[6] (emphasis added)

Physicist David Darling summarized the Copenhagen Interpretation and provided a definition of precisely who can serve as an observer in a statement that brings us very close to the cosmology we will encounter in the book of Genesis:

> *Most astonishingly, in its original form, the Copenhagen interpretation denies that there is any physical reality independent of Man the*

> *Observer*. Our decision to observe, it insists, is crucial, because unless and until we carry out an observation an entity like an electron cannot be said to exist at any place or in any discrete material form at all. In between our attempts to pin it down, it has no existence outside of the probabilistic description given by the wave function. It is not simply that we lack the technical means to keep track of it, instant by instant, or that our theories are not yet precise enough to work out just where the electron is. The fact is, there are no electrons or protons or photons "out there" until we go and look for them. In Bohr's own words: "An independent reality in the ordinary sense can neither be ascribed to the phenomena nor to the agencies of observations."[7] (emphasis added)

America's greatest physicist, John Wheeler, gave the theory of origins that David Darling is describing a very interesting name. Michio Kaku reports that Wheeler boiled it all down to three words and called it "Genesis by Observership."[8] We will look more closely at Wheeler's theoretical work near the end of this book. He offers one particular proposal that is almost identical to Genesis' description of the creation process, a process we need to examine in its entirety in order to appreciate the congruence.

Niels Bohr famously quipped, "Those who are not shocked when they first come across quantum theory cannot possibly have understood it." I suspect the analysis offered here has supported his warning. The big question is, does quantum physics accurately describe the world? We can answer that question because quantum physics has been challenged in every way imaginable and it has met every test. Michio Kaku calls quantum theory "the most successful theory of all time."[9] Let me describe one of these experiments in some detail.

Every material (which means humanly observed) particle has a quality called *spin*. This spin gives each particle a distinctive angle of polarization. Particles can be split in such a way that the two halves of one particle can be thrown toward opposite ends of the universe at the speed of light. Machines have been built that are capable of changing the polarization of one of these split particles and measuring the effect of the change on its partner half particle. Thanks to these devices we discover that when the angle of polarization of one of the two related particles moving apart at the speed of light is changed, its partner instantly "responds" to that change by generating a complementary adjustment. That might not sound important, but it is. This result, confirmed by many experiments going back to the work of Alain Aspect, Jean Dalibard, and Gerard Roger at the Institute of Theoretical and Applied Optics in Paris in 1982, tells us that information is being exchanged between the two half particles even when the particles in question seem to be irretrievably separated and are rushing away from each other at the speed of light. There are only two possible explanations for this. It is possible that some unknown particle is transmitting the information, but if so that completely imaginary particle would be doing so at a speed that would have to be faster than light. This, however, is unlikely for many reasons including the fact that such a particle would, according to relativistic theory, possess infinite mass (it would essentially be a black hole). The other possibility is the idea embraced by quantum physics in its law of nonlocality, a law that states that the two particles act in perfect synchronization because there is no distance separating them at all. The reason there is no distance between them is that the particles are not really physical objects as they seem to be when they are observed. The experiment confirms the proposition that in a normal state of nature the things we call particles are actually what quantum physics says they are: quanta; discrete energy levels that possess an informational rather than material existence. Based on these tested and replicated experiments, John Wheeler suggested that all physical reality may consist of just one electron shuttling back and forth among the various quanta or energy levels, forever weaving the changing cloth of material reality.[10] His statement underscores the point that the universe we live in seems to consist in nonmaterial mathematical information whose physical expression is entirely real but contingent on the presence of a human observer and never more than temporary, lasting only for whatever time that observer is actively taking measurements.

The discoveries of quantum physics are devastating for both evolution and creationism. Stated succinctly, if reality normally takes the form of mathematical information (quanta) and particles exist only in the context of a human observer who is actively taking measurements, how can materialism account for the first appearance of the particles such as electrons, protons, mesons, or quarks that provide the foundation for material reality? We need to bear in mind that the Copenhagen Interpretation of quantum physics "denies that there is any physical reality independent of *man the observer*"[11] (emphasis added). Dr. Goswani states the argument this way: "Observers are necessary to bring the universe into being."[12] The materialist assumption that particles can have their own history separate from a human observer is quantum nonsense, and it makes no difference if that imagined history that supposedly preceded the appearance of man stretches over billions of years as evolutionists propose or just the few days allowed by young earth creationists. For evolutionists, the problem is especially vexing. How can material reality evolve the creature whose observations are the very thing that draws mathematical information into material existence? Can a material universe evolve its own progenitor? If particles appear only in the context of a human observer actively taking measurements, materialist philosophy cannot account for the first electron and that means it cannot account for the existence of anything else. In the universe described by quantum physics, any form of philosophical materialism and any kind of materialist science is a non-starter.

If people are so dedicated to the philosophy of materialism that they insist on holding to either evolution or its creationist cousin, they can do so only by abandoning quantum physics. We must again remember that history demonstrates that nothing good happens when we embrace beliefs that can be sustained only by repudiating various laws of physics. Ernst Rutherford had it right when he stated, "All science is either physics or stamp collecting."[13] In any battle between ideology and physics, physics wins. This fact undoubtedly explains why the work of creation/evolutionists (in this context, remember, they are the same thing) is characterized by so much wild speculation and reliance on false evidence. Speculations that are not consistent with the fundamental laws of physics will always fall apart.

Observational Reality

It is easy to move directly from a discussion of quantum physics to the physics of Genesis because the two approaches are basically the same. Genesis always assumed that there was no point in talking about a material universe until a human observer was present and that observer had begun taking measurements, or in Genesis' terminology "naming things." That is exactly the way Genesis tells the story of creation. Prior to the creation of man, the text of Genesis directs our attention to the construction of a schematic or blueprint that defines the informational content that is the enduring form of the universe. As soon as human beings come on the scene and start to make their observations, this blueprint will, in Genesis, take physical form, but not before that event. In Genesis, Moses answered the questions philosophical materialism cannot even pose: Where does the information that orders our world come from? And what is the content of that information? In doing that, Moses told us what we *really* need to know about our world.

The most important lesson Genesis teaches is that the universe existed in the mind of God as a blueprint or schematic arranged in the form of a sequence of digits before it ever took material shape. Physicist James Jeans unintentionally summarized the message of Genesis and perhaps of the entire Bible when he suggested that "The Great Architect seems to be a mathematician."[14]

One expression of the idea that the universe initially existed as a mathematically organized schematic appears in Proverbs 8. Proverbs is a book celebrating *hokma,* a Hebrew term often translated "wisdom" but a term that includes mathematical and scientific knowledge as well. In Proverbs, the mathematical structure of the universe, what we call the information that defines existence, is personified as a woman. This is what Proverbs has to say about that woman (Lady Wisdom) and the ideas she represents:

> The Lord possessed me at the beginning of His Way
> Before His works of old.
> I have been established from the beginning, before there ever was an earth...
> Then I was beside Him as a master craftsman
> And I was daily His delight...
> Now therefore listen to me my children,
> For blessed are those who keep my ways.
> Hear instruction and be wise, and do not disdain it.
>
> —Prov. 8:22–23, 30, 32–33

This passage tells us that before anything material came into existence, laws of physics and morality were already established, laws that contained information that would eventually give the physical world its structure. Proverbs pictures creation not as a manufacturing operation in which objects are the end product but as a process similar to writing a computer program in which the result is a code containing a series of commands. That code, once written, is what the Bible calls wisdom or *hokma*.

In John 1, we are offered another description of the creation of the universe. This description is very similar to Genesis, but approaches the topic in a slightly different way. According to St. John, this is how creation occurred:

> In the beginning was the Word, and the Word was with God and the Word was God. He was in the beginning with God. All things were made through Him and without Him nothing was made that was made.
>
> —John 1:1–2

The term used here, *Word*, renders the Greek noun *logos*, which refers to anything that has rational content. In that sense it expresses the same idea as the English term *information*. It also expresses equally the concept of word and the concept of number, for in Greek every letter and therefore every word communicates a number. The notion here is that before any material thing came to be, numbers existed, numbers were with God (echoing Proverbs), and the essence of creation consisted in the arranging of numbers. With this in mind, the text confirms that nothing physical was made until the numbers that are the software program of creation were in place.

One more text deserves special attention. In the first chapter of his Letter to the Romans, a part of the New Testament, St. Paul writes:

> For the wrath of God is revealed from heaven against all ungodliness and unrighteousness of men who suppress the truth in unrighteousness, because what may be known of God is manifest in them, for God has shown it to them. For since the creation of the world His invisible attributes are clearly seen, being understood by the things that are made, even His eternal power and Godhead, so that they are without excuse because, although they knew God, they did not glorify Him as God, nor were they thankful, but became futile in their thoughts, and their foolish hearts were darkened. Professing to be wise, they became fools.
>
> —Rom. 1:18–22

This remarkable paragraph asserts that deep in their hearts everybody knows that God created this universe. Some people reject that idea but they do not do so based on any evidence that the claim is untrue (there is no such evidence), they do so based on personal preference. This text tells us that such people are not so much bothered by the notion that there is a Creator as they are by the moral claims that acknowledgment of the existence of a Creator might place on their lives. The idea that there might be a Creator opens the possibility that there could be such things as sins and along with them a need for people to repent and seek forgiveness.

The most interesting words in this paragraph are the words referring to God's "eternal power," which Paul insists is visible in nature. The words *eternal power* render the Greek term *aidos*, a word that at its root *(aei)* refers to "recurring patterns." The idea is that nature exhibits repeating patterns, patterns that reflect the essence of God's creative work. The text tells us that while these patterns are in themselves nonmaterial, they are nevertheless discernable, which means they can be collapsed into a material form, observed, and measured. This passage once again confirms the idea that the material world we see is the reflection of a schematic or blueprint that had prior existence both in time and in importance. As an aside, I want to offer a conjecture that Paul here is referring to the kind of archetypal reiterations observed and described by chaos theory.

It is worthwhile to spend a moment looking at chaos theory. While it is clear that the greatest danger to materialist philosophy and science comes from quantum physics, other nonmaterialist approaches pose similar problems. The challenge posed by chaos theory is particularly interesting for Darwinists. Before chaos theory came on the scene, Darwinists could with some justification argue that the similarity of all forms of DNA might indicate descent from a common ancestor. But chaos theory offers an alternative explanation for this similarity, an explanation that has a big advantage in that, unlike the Darwinian conjecture, it can be tested. (Speculations about events that supposedly happened once in the distant past are, by definition, not subject to experiment or falsification and are not in that sense truly scientific.)

The Strange Attractor

According to chaos theory, phenomena tend to reiterate certain basic structural forms. This is so because nature (and we might as well throw in culture as well) turns out to be stereotypical and archetypal in the way it organizes itself. Similar structures, like the shape of a fern leaf, recur throughout nature at macro and micro levels. These same patterns can even be generated by computers grinding out number sequences when those sequences are translated into visible patterns on a screen. What this means is that the presence of common morphologies, such as similarities in DNA structure, can never prove a common history. The reiterations we see in nature and culture may be nothing more than products of a force that chaos theorists call, for lack of a better term, the *strange attractor*. It may turn out that all things are related but not in a linear sequence of the sort evolutionists conjecture. Things may be related by an unknown *nonmaterial* structuring factor located in mathematics that determines the form of the material expression of everything. We might say that things are similar because reality is everywhere the product of the same software program. The main point is this: if science accepts even the vaguest possibility that archetypal structures might exist, from that point on there is no way to discern whether a correspondence in the appearance or operation of two entities is a result of diffusion (a shared history) or the reiteration of a universal archetype.

I do not know with certainty that St. Paul was referring directly to the kind of reiterations chaos theory studies, but in the passage from Romans just quoted, his basic point is clear. The universe presents orderly patterns that testify to a source whose thoughts are reflected in those patterns, patterns that provide the information that can, under conditions of human observation and measurement, temporarily assume material form.

The theme that unites all these texts is the fact that they all present the act of Creation in distinctly nonmaterial terms. The creation these verses describe involves the generation of *hokma*, *logos*, and *aidos*, all terms referring to what we would call a blueprint. Nothing in these pivotal verses gives us warrant to think of the act of Creation as directly involving the construction of any material entities. These texts support our thesis that reading Genesis is like pressing the "reveal codes" key on a computer; it lets us see what is going on behind the windows.

Verse two of Genesis 1 continues to reveal those answers: "The earth was without form, and void, and darkness was on the face of the deep. And the Spirit of God was hovering over the face of the waters" (Gen. 1:2).

This verse is interesting because there are two ways to interpret it. Some exegetes propose that the language of the verse might hint at a "gap" between the first and second verses in Genesis, a gap that might, they say, hide billions of years of lost history. For them, the reference to chaotic conditions refers to a loss of integrity that entered the creation at some point following its formation. It is obvious that this speculation is nothing more than an attempt to harmonize Genesis with the materialist notion of vast geological ages. It is equally obvious that, as the old phrase has it, that dog just won't hunt. The problem is that there is no support anywhere else in the Bible for such a reading. The Bible is an integrated message, consistent in its language and themes, and the Bible teaches throughout that God created the universe just once and did so in six days. There is no reference to God starting creation, not liking what He saw, taking a mulligan, and beginning again.

The traditional interpretation of the second verse is actually a lot more compelling than any revision. It is noteworthy that, in Hebrew, it is possible to read Genesis 1:1–2 as a single sentence. That sentence might read, "*When* God began to create the heavens and the earth, the earth was without form," etc. This arrangement of the text supports the traditional reading of Genesis 1:2 whereby Genesis is understood to be describing the condition of the primeval universe prior to the beginning of God's work. Enthusiasts of the big bang speculation will notice that "without form," "void," and "darkness" could be read as suggesting the properties of a singularity. The comparison is interesting but I do not believe it grasps the point that Genesis is making.

Genesis is telling us that before God began his work, reality existed in the form of two factors. The first factor was chaos expressed in the language of void, formlessness, and darkness. The second factor is described as the "Spirit of God" or Holy Spirit.

The factor described as the "Spirit of God" is the same strange presence that we have already met in the various passages we cited earlier in this chapter. This is the *logos* or principle of number and logic John spoke about. This is also the *aidos* or "recurring patterns" that St. Paul said are evident to any careful observer who is willing to take an honest look at nature. This is the "master craftsperson" introduced as Lady Wisdom *(hokma)* in the book of Proverbs who describes herself as being present in principle even before God began the work of creation.

In the language of chaos theory, verse two of Genesis Chapter One is describing a state of disequilibrium within which hides a strange attractor with unexpected powers to generate mathematical patterns that are able to organize material structures that take visible form when they are measured by a human observer. In quantum language, this is what physicist David Bohm called the *enfolded order* that lies concealed in the form of pure mathematical information beneath *explicate* (in other words, observed) phenomena. This factor, presented in Genesis under the name "Spirit," is the information we have been looking for. This is the source that is fully able to account for the archetypal reiterations that evolutionists credit to common descent (from what, for them, remains a completely unknown source). This is the blueprint that takes material shape in the form of natural laws that big bang enthusiasts cannot account for at all. This same Spirit is the source of the revelation that gave us the book of Genesis itself.

Light As A Means

Returning to the text of Genesis, the actual creation of the universe is described in the words of verse 3, "Then God said, 'Let there be light' and there was light."

This verse is a favorite among some Christian writers. They sense here a point of contact between big bang speculations and the Bible. The image of light exploding into formless emptiness certainly resembles the language of the big bang theory. But Genesis decisively rejects the premises upon which big bang speculations rest. Working from materialistic presumptions, big bang researchers hope to show that their explosion can generate the entire periodic table out of absolute nothingness. As we have seen, this effort to produce everything out of nothing runs counter to the established principle of science that there is no such thing as spontaneous self-generation. But they face an even more serious challenge. We have seen that contemporary physics views the universe as a system in which information (in the form of probabilities) rather than material objects provides the essential content of existence.

This means that big bang theorists must demonstrate how their conjectured cataclysm generated actual information. Werner Gitt states the challenge well when he notes, "There is no known natural law by which matter can give rise to information, neither is any physical process or material phenomenon known that can do this."[15] Unencumbered by the materialistic obligation to propose a physical source for everything in nature, Genesis does not stumble over either problem. For Genesis, the creation of the universe is preceded and informed by the presence of a Creator whose attributes include *logos* or numerical logic, *aidos* or archetypal patterns, and *hokma* or natural law, all terms that refer directly to information. Because it presumes the existence of such a Creator, Genesis has no problem accounting for the origin of a universe organized and structured by the flow of information, a universe that can be aptly described as a system of communication rather than a collection of objects.

Genesis begins its explication of the structure of existence by directing our attention to the phenomenon of light. The act of creation in Genesis consists in the invention of light. This focus on light reflects the fact that the author of Genesis thinks of the universe in terms of the generation and propagation of information rather than the interaction of objects. Philosopher Ernst Cassirer in his 1946 book *Language and Myth* drew the connection between light and information: "For the mind, only that can be visible which has some definite form; but every form of existence has its *source* in some peculiar way of seeing"[16] (emphasis added).

I will be saying several things about light in this chapter and the next. But the first point I should make is that light exists in two forms, a physical form called a *photon* and a wave form in which it presents itself not as a physical entity but as a mathematical probability (the probability refers to a number expressing either its location or direction of movement). This means that light stands at the borderline between pure information (the probability wave) and material existence (the photon). This borderline quality is the aspect of light upon which Genesis Day One focuses. The statement "Let there be light" means, in contemporary physics, "let there be possibilities."

The possibility upon which Genesis centers is the possibility of material existence itself. Light does not guarantee material existence but it does contain its seed. In the next chapter we will see exactly why this is so. For the moment it is enough to say that when we define the universe in terms of information, light plays a determinative role because light is the medium along which information travels. Light also limits the flow of information in ways that give a particular universe its unique structure. The ways in which light defines a universe will also be the subject of the next chapter of this book.

Genesis wants us to re-vision our universe as a system of communication. That is why it starts off with the creation of light, and as we will see, emphasizes the role of light at many places in the text. As I said, light is the medium that makes the communication of information possible. But Genesis goes beyond reporting the creation of the medium of communication. It goes on to describe the content of the information that provides our universe with its structure.

We first glimpse that content in Genesis 1:5. The New King James Bible translates the verse as follows: "And God called the light Day and the darkness He called Night. So the evening and morning were the first day" (Gen. 1:5).

The Hebrew words *day* and *night* are terms that are directly related to the communication of information. In Hebrew, the word *day* is, of course, associated with the diurnal part of a normal earth day. For our purposes, the significant characteristic of daytime is the fact that this is the condition under which things are most readily observed. Daytime is the realm of activity and as such refers to the part of our lives that is involved with information exchange. It is interesting to note that the Hebrew term describing the beginning of a new day literally speaks of something becoming *discernable* or being *revealed*, terms connected with the transmission of information. The Hebrew term connoting *night* renders the idea of gloom, suggesting descent into a silent realm beyond sight and knowledge.

As Different as Night and Day

As we have seen, the Bible understands the words *night* and *day* to literally refer to a twenty-four-hour day. But these terms also tell us that our universe has two distinct aspects. There is a realm whose structure is capable of being measured and known, a part of reality that is made of information. But the same terms tell us that there is another part of existence that is every bit as real but which remains outside the flow of information and, because of that, is forever unknowable to us, a dark world.

This separation of realms anticipates an important idea in quantum physics, the Heisenberg Uncertainty Principle. The principle states that one cannot know both the location and direction of movement of a subatomic particle. The Heisenberg Principle might sound trivial but it is not. Classical physicists believed that if we could for one moment bring together a complete picture of the physical forces operating at one point in time and space, we should, by extrapolation, be able to identify and predict all events going on everywhere in the universe past, present, and future. This was so because classical physics presumed that everything had a knowable cause and measurable effect within a tightly closed circle in which only physical laws and visible objects were real. Classical physics was, like evolutionary biology, a byproduct of philosophical materialism.

The quantum Heisenberg Uncertainty Principle means that the day when we understand every secret will never dawn. The reason is that, according to the principle, the very act of knowing some things prevents us from knowing others. The act of identifying the motion of a particle destroys the capacity to locate it, while locating a particle automatically blocks any ability to measure its motion. Putting that in terms of information, some forms of information cancel access to other information. The universe will always conceal secrets. There will always be a realm of the unknowable or, as Genesis puts it, there will always be "night" and "day." All this implies that at any time our carefully constructed personal worlds can fly loose and head off in just about any direction under the influence of forces we don't know anything about, an important idea within quantum physics that we will explore later. At this point I can't help but mention my all-time favorite scientific wall poster. Capturing the point I am trying to make, it succinctly announces "Heisenberg May Have Slept Here."

In light of the Heisenberg Uncertainty Principle it is interesting to notice that Genesis always lists evening and night first and morning along with day second. That pattern is also encountered in the second verse of Genesis 1 in which the "void" is introduced before the "Spirit," which as we have seen, contains the information that sets the parameters for the visible world. Genesis seems to be suggesting that the darkness is a much larger world than the realm of light/information. If that is so, contemporary physics confirms this insight.

Physicists extrapolate that, at present, we are aware of only about 27 percent of the matter that makes up our universe. The realm of the unknown, what some physicists call "dark matter," makes up 73 percent of the universe. Dark matter exists beyond the realm of light and is, therefore, something other than information. This kind of discovery tells us that undoubtedly reality would look very different if we could peer into the unknown and see what lies beyond even the possibility of vision. We might ask very different questions and set different priorities if we could see reality for what it is. As a revelation from beyond space and time, Genesis will help us do just that.

Numerical Reality

Genesis begins revealing the hidden code that shapes our visible universe by offering two words that end 5:1, words translated in the New King James Version as *first day*. The Hebrew text reads a bit differently; it can more literally be rendered into English as *Day One*. This difference in renderings is highly significant. The correct rendering puts the emphasis on the number rather than the concept of day. In the Bible as a whole, number is a very important idea. Hidden numerical patterns fill the Bible from beginning to end. We will examine some of these in the final chapter, but one example can be discovered early in Genesis 1 itself.

The familiar first verse of Genesis hides a remarkable secret. In Hebrew the verse has seven words and those words include three nouns. In Hebrew, every letter also serves as a number. In this case the letters can be read as numbers totaling 777. Already we can see a pattern in that the entirety of Genesis 1 is built around the number seven as in the reference to seven days of creation. In verse one, the seven words are made of twenty-two letters, the exact number of letters in the Hebrew alphabet. This too is interesting in light of the statement in the gospel of John that "in the beginning was the word and the word was God." The twenty-two letters are the elements out of which all Hebrew words are built. But at this point things get even more interesting. If you divide twenty-two, the number of letters in the first sentence in Genesis, by seven, the number of words in that sentence, the math gives the figure known as "pi" to four digits. Pi expresses the relationship between the radius and circumference of a circle. The concept of pi perfectly fits the definition of information that we established earlier: "Information is neither a physical nor a chemical principle like energy and matter, even though the latter are required as carriers."[17] Pi is not something we can see, it is nonmaterial, but it determines the qualities of the things that we do see. The "coincidence" that Genesis 1 would contain a hidden reference to pi underscores our contention that this text is intent on laying out the *patterns* of creation rather than describing the physical construction of objects.

While Genesis 1:1 is an example of a text that hints at one hidden structuring agent working in the world, the fifth verse of chapter one, with its reference to "Day One," reveals a much more important piece of the information that is shaping the universe. Once again, as a result of many years of study, I have come to the conclusion that Genesis is often misread at this point. In the interpretation of Genesis 1 as a whole, too much emphasis has been placed on the idea of days and not enough on the numbers in the text. I do not argue this to avoid the idea of creation over a six-day period. A six-day creation should not be a problem for anyone interpreting Genesis. For Genesis, the universe is a product of supernatural action. Limits imposed by the assumption that we live without God in a closed universe in which the only reality is matter and the only processes that could generate a universe are physical do not apply to Genesis. But beside that fact, Genesis never intended to offer a description of the physical creation of material objects. Genesis is revealing the thoughts of God that shape the universe we inhabit. It lets us peek at the schematic that is the informational source from which we construct the material world by way of our measurements and observations. Genesis does not need to explain the formation of material entities because once the blueprint is written and the rules governing the game of life are set in place, all that remains for God is to design creatures capable of collapsing the code. Those creatures will reduce the numerical archetypes that flow from the mind of God to material entities through their acts of observation and measurement.

The first archetype revealed in Genesis is the number one. To understand the importance of one in Genesis we do well to return to our earlier study of Pythagorean thought. I have noted that the ancients widely believed that Pythagoras based his understanding of numbers on the writings of Moses. I want to accept that as a working proposition and see where it leads us.

In Pythagorean theory, the digits "one" and "two" are concepts but they are not numbers. The first true number is always "three." In the view of Pythagoras, "one" is an archetype prior to all numbers. As Charles Kahn puts it in his book, *Pythagoras and the Pythagoreans*, "Cosmogony begins as the numbers are generated … thus the cosmos arises from *the One* by breathing."[18] It is important to remember that for Pythagoreans material existence is simply the outer husk of a reality whose substance is numbers and that the study of the universe demands that we study those numbers. Both Platonists and Pythagoreans argued, in fact, that the study of material entities apart from number will always prove misleading.[19]

In Pythagorean theory, the reason that "one" is the beginning of all things is geometrical, not sequential. In geometrical terms one is represented by the point. "Two" (again, the digit was not considered a number) extends that point and turns it into a line. From the line, all the shapes of plane geometry are generated by the addition of points (numbers) and lines, and a physical universe becomes a possibility. As we will see in the next chapter, this is precisely what happens in Genesis on Day Two of creation with the formation of the "firmament," which in actuality is defined in Genesis as a solid linear expanse (that is, as a line). The basic idea is that for anything to exist there must first be a point and the idea of the point is expressed by the number one.

So who, or what, does the number one in Genesis refer to? Who, or what, makes our information-rich reality possible according to Genesis? It might at first glance seem likely that the "One" in Genesis would refer to the Creator himself. But I do not believe that is the case for the simple reason that the Creator is not, in Genesis or in the Bible as a whole, a part of His creation. The biblical God is beyond space and time, whereas the number one is a form-giving structure that operates upon creation from within the creation.

When we examine the text of Genesis we can clearly see that the number one refers back to the word *day*, not God. The text, after all, speaks of Day One. That is, where the link is made, not between one and God. But to what, then, does day refer? We have seen that in Hebrew the word *day* refers to the realm in which information flows as opposed to night when nothing is seen or known. The concept "one" should, therefore, be understood to refer back, in Genesis, to the idea of light.

As we have seen, light is the medium that makes the communication of information possible. It is, therefore, the force that gives the universe (which is a system of communication and not a collection of objects) its existence. Without light, information would still exist in the mind of God but it would not be communicated in such a way as to generate a universe that reflects the mind of its Creator. Without light, there would be no conscious entities capable of searching the structures of existence for clues into the thinking of their Creator, for as we will see in the next chapter, the very existence of conscious awareness is predicated on certain qualities inherent in light.

For all these reasons, it is best to read the reference to "one" in Genesis 1:5 as the revelation of an archetype present within the substance of Creation and not as a reference to the Creator himself. The concept "one" in Genesis must then refer to the organizing principle of light. This is important because with this verse a pattern emerges that guides the interpretation of the rest of Genesis 1. As we move through the text we will see that each number associated with each day can be read as an archetype out of which some aspect of our universe arises, an aspect associated with the themes of that day. The focus of Genesis will remain steady on the generation of these archetypes and not on the origin of any materiality that is mentioned in connection with the archetype. The text of Genesis will not waver from its purpose, which is to describe the creation of the universe as the encoding of a field of information.

Chapter 5

Setting the Limits of Reality: Day Two of Genesis

The second day of creation is described in only three verses, but for any effort to understand the cosmology of the Bible these may be some of the most important verses in Genesis. They have also occasioned controversy and confusion.

The interpretation of Genesis, Day Two, rests on the assumptions one makes about what Genesis is telling us. In this book, I am proposing that Genesis has little to say about the creation of the material world as such. Rather, Genesis is offering a peek behind the appearances of material reality that allows us to see how the universe is organized. Genesis does not offer a history of the physical world. It does not do that because it does not operate within a materialist premise. In Genesis, nature is viewed as a field of information. Genesis wants us to know the numerical codes and geometric principles that give material reality its content. It reveals the *logos*, the *aidos* or divine pattern, and the *hokma* of the creation.

If we can grasp the message of Genesis we gain knowledge far more useful than anything materialism can offer. Material objects come and go. The prophet Isaiah underscored that fact when he wrote, "The grass withers, the flower fades" (Isa. 40:8a). Grass and flowers are material entities. They last exactly as long as the attention of the observer whose measurements conjure their existence. Being made of electrons and protons, these things possess an ephemeral existence well described by the mathematician/philosopher James Jeans when he wrote, "It is probably as meaningless to discuss how much room an electron takes up as it is to discuss how much room a fear, an anxiety, or an uncertainty takes up."[1] But the same Isaiah who spoke of the ephemeral existence of material things also affirmed that "the word of our God stands forever" (Isa. 40:8b). The deep archetypes, the word or *logos* of God that define reality, never change. Each new arrangement of physical things will always conform to the Divine patterns alluded to in Romans 1. This means, as James Jeans suggests, that the effort of so many to understand material objects may be an interesting hobby, but it can never be anything more than that. The crucial question is, can we grasp the substance behind the appearances, the blueprint that provides the foundation for everything we see? To use Einstein's favorite description of the ultimate goal of all research, can we glimpse the "mind of God"?

Day Two describes the creation of something called the *firmament*. Here is what it says:

> Then God said, "Let there be a firmament in the midst of the waters, and let it divide the waters from the waters. Thus God made the firmament, and divided the waters which were under the firmament from the waters which were above the firmament and it was so. And God called the firmament Heaven. So the evening and the morning were the second day.
>
> —Gen. 1:6–8

A glance at almost any Bible dictionary will yield the information that the Hebrew word rendered as firmament means "sky." This definition is usually accompanied by a diagram that explains the Bible's purported "three-level view of the universe," a picture featuring heaven above and hell below with earth in the middle separated from heaven by an empty space labeled "the firmament." The first thing we need to do is toss that picture.

The word *firmament*, at its root, has nothing to do with the word *sky*. The term literally means "an expanse," nothing more. The reading that finds in that word a reference to sky is an interpretative extension. It is a popular rendering, but one that quickly runs afoul of the actual usage in the Genesis passage.

Separate and Unequal

In Genesis, the first thing that is said about the firmament is that it divides. In Genesis, firmament is something that divides and that is all that it does. It is not some sort of roof over the world (a world that, in the text, does not yet exist). The second clue into the meaning of *firmament* comes when we are told exactly what the firmament divides. In Genesis, firmament divides water from water. Note well, firmament does not divide earth from heaven, or land from space. It is an expanse that separates two areas of water, one below and one above.

If we review the use of the term throughout the Bible, it is instructive to note that scripture itself defines *firmament* in an unexpected way. In verse 8 we are told that "God called the firmament Heaven." In the classic diagrams we spoke of, heaven is usually pictured as a place sitting *above* the firmament. That is not what Genesis says. For Genesis, the firmament is itself the residence of God. It is, of course, commonplace to say that God lives in the sky, which is one reason that *firmament* is sometimes used as a metaphor referring to the sky. But remember, we are examining a word that literally means nothing more than "expanse," a word that indicates something that divides waters from waters, not earth from space. In such a situation, the reference to heaven, far from establishing a connection between sky and firmament, actually confirms our argument that the biblical term *firmament* refers to something other than the sky. Heaven, in the Bible, is a more interesting and complex reality than the sky.

We gain further clues into the meaning of the term by looking at ways in which it is used in the Bible outside the book of Genesis. In Ezekiel 1:22, the prophet is given a vision. The vision is an odd one if *firmament* only means "sky" because the revelation he is given is a vision of the "firmament" itself. One hardly needs divine help to see the sky (except maybe in a Willamette Valley winter in Oregon). In this vision he describes "the likeness of the firmament" as the "color of the terrible crystal." I have no idea what that means, but I can say with some confidence that the prophet does not seem to be describing the sky.

Throughout the Bible, odd terms surround the word *firmament*, reinforcing the likelihood that the word refers to something more interesting than sky. Psalm 150 speaks of "the firmament of [God's] power." The Psalm asks that praises to God be offered in that place, indicating that it is inhabited. In chapter twelve of his book, Daniel speaks of the "brightness of the firmament," indicating an association between firmament and the continuous presence of light.

In other biblical references, the term *sky* would seem a proper reading for the word rendered as *firmament*. For example, Genesis 1:14 describes lights that are hung in the firmament, a reference to sun, moon, and stars. Genesis 1:20 depicts birds flying in the open firmament above the earth. We will examine these verses later in this study of Genesis. For the moment it is worth remembering that in all languages words often have an exact use and a poetic use. A good example is the English word *heaven*. In precise terms, the word refers to the abode of God. But we all know that it is extended in a variety of casual ways. To this day, heaven can describe the night sky. It can also describe a beautiful view or a wonderful day with a friend. Similarly, the term *evolution* has a precise meaning for Darwinists but it is often used to describe changes that have nothing to do with Darwinian theory, as when a young bride says that her relationship with her husband is "evolving." Language is full of precisely defined technical terms subject to metaphorical extensions in popular usage; for example, quantum leap, black hole, and nuclear power as in "nuke."

To understand the meaning of the term *firmament* as used in the description of Genesis Day Two, we must remember our analysis of Genesis, which sees it as presenting the blueprint or software code of existence. That means we need to focus on the number two. To do that, it is helpful to return to the world of Pythagorean thought, recalling the claim that the Pythagorean tradition finds its roots in the writings of Moses.

Pythagoras systematically described the world in terms of numbers. The neo-Platonist philosopher Nicomachus described Pythagoras's approach this way: "Arithmetic itself preexists in the thought of the artisan God before everything else as a paradigmatic ordering plan (*logos*) on which the craftsman of the universe relies as on a design and archetype for ordering the things produced from matter and making them achieve their proper end."[2]

In Pythagorean thought, the digit two is the Dyad. Like one, it is not actually a number; the first number is three. The Dyad, however, is the precursor that makes numbers possible. Dyad is the principle of distinction and separation. It is the factor that allows us to say that a certain thing is this and not that. As such, it is basically the same as consciousness.

Conscious Knowing

Consciousness is the awareness of distinction. To be conscious of something is to be aware that this thing is not that thing. Consciousness is always a matter of separation, actually two kinds of separation. There is, first of all, the awareness of separation between the observer and the thing observed. Consciousness implies that we perceive something as "out there." We sense that it is different from what I perceive "in here," or in other words, it is not me. The inability to distinguish self and other represents a breakdown in consciousness as in the case of a psychotic state. Such a state is either unconscious or preconscious.

Consciousness also involves distinction between two objects in the external world. We cannot claim consciousness of a thing until we can identify that thing, name it, and understand that it is not the same as other things.

It is important that we accurately define consciousness because we now understand that the world we live in is shaped, at least its material manifestation, by the kind of awareness we bring to that world. This is the principle enshrined in quantum physics as the law of nonseparability or the observer effect. There is simply no way to talk about the universe as such. The only universe we know is the universe we observe. That same universe is also the only universe about which we can communicate information to others. This means that for us there can be no other cosmos beside the universe that we receive and share as information. It does not matter how real a thing is, if we have not established consciousness of it and named it, for us, it just isn't there and we can't talk about it. The core idea here is that for a thing to exist (for us, not for God) it must exist in the form of information that we have in some way observed, measured, and named. This means that in our experiential universe, the universe itself, information, and conscious awareness are all the same thing. Physicist John Wheeler described this relationship between consciousness and existence as a "meaning circuit," a term that Amit Goswami nicely defines in the phrase, "I am here because of you and you are here because of me." [3]

When, therefore, Genesis insists that the Dyad must exist before any numbers (primal information) and any material thing that arises from the collapsing of the primal information can appear, it is telling us that consciousness must precede existence. To put this in philosophical terms, epistemology (the science of how we know something) is primary to ontology (the study of being itself). In the quantum understanding of the material world, being flows from knowing; it is not the other way around. The materialist believes that things exist as physical entities. He studies them and so reduces them (he thinks) to information (notes, pictures, memories, etc.). In reality, all things permanently exist as information that we temporarily collapse for purposes of analysis but that quickly return to their natural state as information as soon as we withdraw our attention. This great reversal of philosophical tradition (a reversal first described by Kant and Hume) stands at the core of emerging approaches to scientific work in every discipline. It also stands at the core of Genesis' cosmology in which the "number" two (expressing the principle of separation or conscious awareness) must come into existence prior to the establishment of any of the archetypes (3–6) that make a material universe possible.

But there is more to the Dyad than the idea of consciousness, as important as that is. As we have seen, the concept (not number) one is expressed in Pythagorean geometry as a point. The point is important but it communicates no real information by itself. A point in space is a singularity. Without a reference to compare and contrast it, such a point conveys zero data. The point only begins to communicate information when it is joined via the dyad to another point. This joining forms a line. The line establishes a plane and as such lays the foundation for plane geometry. The line is not geometry itself but it makes geometry possible. The Dyad represents both the principle of consciousness and the foundation of geometry. In Pythagorean thought, these two ideas are linked because in Pythagorean analysis geometry is the connecting point between consciousness and material existence. In Pythagorean philosophy, the universe begins as a mathematical thought but that thought is transmitted into material form via the principles of geometry. Likewise, in contemporary physics and in Genesis the universe begins and normally exists as discrete nonmaterial energy states (quanta) that are then translated into material objects by the operations of consciousness through observation and measurement or, as Genesis puts it, by naming. The Pythagorean and quantum/Genesis positions are remarkably similar.

It is easy to see how ancient historians could have credited the origin of Pythagorean thought to Genesis. In Genesis, formation of a material universe is preceded by the establishment of the principle of the Dyad. This Dyad is described as "an expanse." The expanse in question is clearly not the sky (except by poetic extension); it is simply a line. But how does this line correspond to the idea of conscious awareness? To be more precise, how does this line, in the teaching of Genesis, provide the precondition for the appearance of the kind of conscious awareness that transforms a universe made of information into the myriad physical entities that we see everyday? The answer is to be found by returning to the topic of light.

The Paradox of Light

The most important thing to know about light is that it is a limiting factor, or, better yet, *the* limiting factor in the propagation of information. Many are familiar with one of the ways in which light limits the flow of information. My wife's car boasts a bumper sticker that announces "186,000 miles per second, It's not just a good idea, it's the law!" Light sets the speed limit of the universe, 186,000 miles per second. It is interesting to note that the speed of light is a formal definition maintained as a scientific convention. As actually observed, the speed of light has been shown to vary widely in different environments. Under controlled conditions in laboratories, light has been slowed to less than four miles per hour. But none of these variances are significant. There is nothing magic about the speed of light because it is not the speed that makes light important. The thing that makes light the limit it is rests in the fact that in whatever environment it operates, and at whatever speed it moves, light is nature's information carrier.

In our universe, the natural state of any entity is the form of a quantum, a mathematically predetermined energy level. To put that another way, the natural form of every person, creature, and object is information. It follows that the shape of our universe will be determined in large measure by the quality of the carrier that communicates all that information. That carrier is light. Since the universe is given its content by the information that is communicated within that universe, every quality of our universe is a byproduct of the capacities of light. Time, for example, is a byproduct of the speed at which information is moving in the environment an observer inhabits. Distance is, similarly, an illusion produced by the speed at which information is flowing. The speed of light served as Einstein's mathematical constant not because it was a physical barrier but because the speed at which light and, therefore, information is moving defines every quality expressed by our universe. This is so because the universe is a communication system and in any communication system information, the conscious awareness of information, and the universe or system itself are just different names for the same reality.

So far we have been speaking about light only in connection with its speed or, to put that another way, the rate of information transfer. The ways in which the speed of light limits reality for us are familiar to many people, even if all the ramifications are not. But light plays a role in another limiting factor that has nothing to do with speed, a limit that is less well known but that is more important than speed. Light helps determine the smallest unit of length possible by which the universe can be measured, a limit that is related to our activities as observers.

Recalling our earlier discussion, the Heisenberg Uncertainty Principle tells us that it is impossible to measure both the location and motion of a particle at the same time. This is so because all things normally exist as energy states and not particles. If we were to supply electromagnetic energy sufficient to enable us to accurately identify both the location and speed of a particle, the energy produced by that observation would generate a black hole. A black hole is an entity that emits no light and, therefore by definition, generates no information. Since entities do not exist for us unless they exist as information, this means that *for us* (not for God) there is a minimum length of motion we can measure because below that length motion and location would, in fact, be the same thing. In physics, this smallest measurable unit is expressed as the Planck length, an indivisible unit of measurement, 10–33 centimeters. It is interesting to note that by the same logic there is necessarily a smallest unit by which time can be measured, the time needed for a particle to travel the Planck length at the speed of light. I will discuss that time limitation later in this book.

The thing that makes the Planck length important for understanding Genesis is the fact that the real universe seems to contain many more dimensions than we experience in the bubble-universe or mini-universe that we are able to perceive. In terms of dimensions, we are aware of three spatial dimensions plus time. Because of Einstein's work, we now understand time to be just another spatial dimension. Like the other three spatial dimensions, time is subject to gravity, mass, and acceleration. Experiments with atomic clocks have demonstrated beyond question that time moves at different rates at different altitudes (gravity is a factor) and varies with one's speed of travel (acceleration and velocity are factors). Since Einstein we have, therefore, understood our universe to consist of four spatial dimensions or more accurately as three-and-a-half dimensions since there are many oddities in our experience of the fourth physical dimension.

While this four dimensional quality may describe the universe we experience, it does not seem to describe the larger universe beyond our observations. The mathematics of quantum physics suggests that the real universe may have at least ten dimensions, not four. It is interesting to note, purely as a sideline, that the first suggestion that reality consists of ten dimensions was offered by Rabbi Moses ben Nachman of Gerona in 1194. Rabbi Nachman based his conclusion entirely on his study of Genesis 1.[4]

The question we might ask is this: If quantum calculations insist that there are, indeed, ten dimensions in the real universe, where are the six dimensions of which we have no visible clue? Current research indicates that they may be folded in under the Planck length. This does not mean that these dimensions are tiny. Size here is not the issue. Remember, the units that we are describing are energy states, not physical objects. The Planck length is an informational barrier, not a physical limitation. Anything existing within the Planck length is not part of our universe only because any information there cannot be accessed by an observer. Since the Planck length is an energy level, whole universes much larger than our own could hide behind this limitation, but our problem is that we can't access them as information. Anything that cannot become part of human consciousness cannot be part of our universe since the universe we inhabit is coterminous with our consciousness.

The Observer and the Observed

Genesis tells us that creation began when God (source of number, word, logic, and structure; see Day One) said, "Let there be light." The invention of light meant that, limited by the Planck length, the universe He created would be a pocket universe or bubble-verse consisting of only four dimensions rather than the full ten. It meant that it would have peculiar qualities related to time and spatial relations that would be defined by the speed of light. This also meant that the universe so defined would be intelligible. Intelligibility would be ensured because the bubble-verse would exhibit two kinds of separation. First, it would be divided into two parts, that which *is* (from the point of view of the observer within the created bubble) and that which is *not* (from the observer's point of view). Second, once collapsed into material form, it would manifest itself as something separate from the observer who collapsed the wave function and generated the visible universe. Mathematician G. Spencer Brown described that second subject/object separation: "We cannot escape the fact that the world we know is constructed in order (and thus in such a way as to be able) to see itself ... but in order to do so, evidently it must first cut itself up into at least one state which sees, and at least one other state which is seen."[5]

These separations are, as we have seen, the necessary precondition for the existence of consciousness. For conscious awareness to exist, a universe must express a limit in the form of a distinction between that which "is" and that which "is not" and there must be at least the pretense of a difference between the observer and what s/he observes. The message of Genesis 1:6–8 is that on the second day of creation God established the Dyad, and it is from the Dyad that both the possibility of information and the possibility for a consciousness that could receive information and collapse it into material objects simultaneously arose.

Once more, we need to remind ourselves that none of this activity created anything in the sense favored by materialist philosophy. Rather, what God has accomplished to this point is the construction of a field in which information can potentially exist and the establishment of the precondition for the emergence of an entity that can translate that information into physical objects. There is division, there is something that is and something that is not, and there is the potential for consciousness or the awareness of dyadic distinction, the capacity to distinguish self and other. That is all that there is but in a universe defined by the communication of information that is also all that is needed. Materiality will immediately emerge whenever it suits God to design a human observer and whenever he or she begins taking measurements or in the language of Genesis, naming things. The Creator ends Day Two at that point.

Our Heavenly Abode

Beside the major themes of the day, the Day Two narrative offers two surprises worth noting. The first surprise concerns the world "firmament." As we have seen, in various Bible passages the firmament itself is called "heaven." In this regard, I want to offer a personal suggestion. You can take it or leave it. Believers often think of heaven, the abode of God, as a place that is above and at some distance from us. It is by metaphorical extension of this common idea that the sky comes to bear the name of heaven, a popular metaphor adopted even within the Bible. But in the precise language of Genesis, heaven is located not up but within, within the line that is called the firmament. It is a basic rule of biblical interpretation that the first use of a term goes a long way to defining that term. If that is the case, many believers may want to redraw their picture of heaven because the first use of the term *heaven* clearly locates it *within* the line of the firmament.

It is entirely possible that heaven is not someplace far away but is right here, woven into the warp and woof of existence as we know it. If we have read Genesis correctly, heaven exists at the border between the four spatial dimensions we experience and the six we can only infer, the place called firmament in the Bible, which is also the extension we have identified as the Planck length. It may be that the throne of God is and always has been set in our midst, but at a boundary defined by light within a larger universe in which we participate only in part. Such an understanding would place God close at hand in exactly the manner that the Bible describes. St. Paul tells us, "We live and move and have our being in God" (Acts 17:18) and Jesus promised us that even while He was sitting at the right hand of God in heaven he would still be "with you always, even to the end of the age" (Matt. 28:20). If it lies within the firmament as we have defined it, heaven sits at the borderlands between the visible and invisible, the eternal and the ephemeral. In personal terms, it stands at the boundary between conscious awareness and the endless depths of the unconscious. Like the red thread traditionally woven through every strand of a Navajo rug, heaven defines the borders of time and space without participating in the limits of either.

If heaven lies within the firmament, it is easy to see how God can be both immanent and transcendent, part of our daily life and yet completely separate from this world. God can do this because He is present in every atom of our universe at the border (firmament) that defines the parameters of observable reality. And yet He is sufficiently separate from the universe so as not to be subject to collapse into visible form because of a desire on our part to take His measure. God can and does take visible form, but only in accordance with His will, not ours.

It also becomes easy to resolve the ancient question that asks how God can know all that is to happen and yet hold us responsible for the content and outcome of our deeds. The answer lies in the flow of information. The Bible tells us that God alone knows "the end from the beginning" (Isa. 46:10). God sits entirely outside the flow of information as we experience it and yet His life flows through that information and He is part of it all. We, however, do not stand outside the flow of information and are, therefore, responsible for how we choose to deal with the challenges of reality as we encounter them. The fact that we experience the world inaccurately as a sequence of physical causes and material effects does not provide an excuse for failure on our part to make the right choices. Quite the contrary, our universe seems to be specifically structured to facilitate our growth in conscious awareness by giving us a chance to make moral choices within a field of artificially limited options.

We have to consider the possibility that this universe we inhabit was generated for pedagogical purposes almost like a virtual reality program. The Bible insists that God is showing his love precisely by challenging and testing us. In Revelation 3:19, God tells us,

"As many as I love, I rebuke and chasten. Therefore be zealous and repent." I have a hunch that the trials and temptations that afflict us hide the very purpose of our lives, which is fulfilled in the dawning of our awareness of a need for a savior and in a growing hunger for the wisdom that flows from God's Word.

This brings us to the other surprise found in the Day Two narrative. Day Two is the only day about which God does not say, "And it was good." At the end of Day Two, God is silent. To recount the speculations offered over the centuries about this odd silence would take us far afield, but I do not think we need to make the problem all that complicated. God did not pronounce a verdict on His work of separation at the time of creation because there was no verdict to pronounce. Separation is just that, separation. It generates possibilities and consciousness, nothing more. The possibilities that will be actualized in the choices we make will determine the goodness of what has been accomplished. Conscious awareness can turn out to be a blessing or a curse. No one can render a verdict on consciousness per se. A verdict can only be rendered on the lives and the world that consciousness carves out of the chaos of the void, and the Bible assures us that verdict will one day be rendered. But the Day Two verdict will be announced at the end of the story before the Great White Throne set in the midst of a creation restored to fellowship with its Creator, not before.

Chapter 6

The Mystery of Three: Unlocking the Secret of Life

We are exploring the creation of the universe from a perspective outside the boundaries of time and space. The biblical book of Genesis is offering us a picture of our world that is radically different from anything available in any science textbook. The reason it is different is a matter of perspective.

We have grown accustomed to looking at the world from many different perspectives. We have, for example, become familiar with the way our world looks from low earth orbit, a view denied mankind for most of its history. In Genesis, the perspective is even more grand. We are looking at our entire universe from the perspective of its Creator. Talk about a view from above! We are seeing our universe from heaven's mezzanine.

Genesis describes our universe as a system of communication rather than a collection of objects. Because the universe is a system of communication, the act of creation is not a story about the construction of creatures. Creation takes place through the generation of information and the invention of a medium to propagate that information. In Genesis that medium was established on Day One when God said, "Let there be light." Following Day One, Genesis began to reveal the hidden codes around which the universe is organized. Those codes are presented as a series of numbers, numbers that contain the seed of everything that will exist in the material universe when it comes to expression under the eye of human observation and measurement. For Genesis and its heavenly perspective, these numbers are the most important things to know for they constitute *logos, hokma,* and *aidos*, the rational structure and eternal pattern that will shape all things. Compared to numbers, materiality is a derivative and artificial world.

In Genesis so far we have met two archetypes associated with the digits one and two. We have used the Pythagorean tradition as a guide to understanding these archetypes based on the ancient claim that Pythagorean thinking finds its root in Genesis. In Pythagorean mathematics, one and two are not numbers, they are concepts. As such, they have no direct relationship to any material reality. They make possible the existence of the information flow that can take shape as materiality in the presence of a human observer actively taking measurements, but they are not themselves information. We have seen that Genesis treats these two archetypes in the Pythagorean manner as precursors to the flow of information. In Genesis, the concepts one and two both involve light, first its appearance and then the formation, by means of light, of a division between known and unknown universes and between self and other. Light makes possible the conveyance of information and division creates the possibility of consciousness that can receive and use that information.

Information actually begins to flow in Genesis with the emergence of the first true number. That number is three. In Pythagorean thought (and Platonic thought as well), three automatically appears when one and two are established. This is a matter of simple arithmetic: if there is a one and a two there is in that very fact also a three. To put that in informational terms, once the conditions for the flow of information are in place, information starts to fill the established channel.

As we have seen, in Pythagorean thought, one generates the point and two creates the line. Point and line are the preconditions for geometry, the sacred science for Pythagoreans, but they are not yet geometry. Geometry begins with the first true number, three. This is so because three adds the crucial third line necessary to create a solid figure.

In Pythagorean thought, this number three is expressed by the all important form of the triangle. Anyone familiar with Pythagorean geometry will remember that for that school the triangle was a matter of utmost importance and intense study. The famous Pythagorean theorem has to do with the angles of a hypotenuse triangle.

But there was more to the number three than its association with the triangle. For Pythagoreans, the triangle itself was important because all the three-dimensional Platonic solids were constructed on the basis of the triangle. In Plato's *Timeaus*, a work deeply influenced by the Pythagorean tradition, we are told that the elementary bodies that make up the world are constructed "out of two right triangles that combine to form the five regular solids."[1] These solids were understood to provide the unseen link that made it theoretically possible for numbers to take expression as materiality. It is therefore accurate to say that the potential existence of all animate and inanimate material objects rests on the archetypal structure called the number three because three generates the triangle and from the triangle the Platonic solids are formed.

This is precisely the way in which the number three is introduced in the book of Genesis, a writing that by ancient tradition provided the template for Pythagoras's ideas. On Day Three in Genesis, the creation literally explodes into existence. This explosion includes the establishment of the mathematical foundation for both the animate and inanimate worlds. Where the day before there was nothing but point and line, suddenly like a building whose foundations seem to sprout walls one Tuesday morning, there is a lot of stuff there! Let's look at the text:

> Then God said, "Let the waters under the heavens be gathered together into one place, and let the dry land appear;" and it was so. And God called the dry land Earth, and the gathering together of the waters He called Seas. And God saw that it was good. Then God said, "Let the earth bring forth grass, the herb

> that yields seed, and the fruit tree that yields fruit according to its kind, whose seed is in itself, on the earth," and it was so. And the earth brought forth grass, the herb that yields seed according to its kind, and the tree that yields fruit, whose seed is in itself according to its kind. And God saw that it was good. So the evening and the morning were the third day.
>
> —Gen. 1:9–13

It is important to carefully examine the flow and internal logic of this text. The text begins with a reference to waters because "waters" is all there is at this point. As we have seen, the term *waters* in Genesis refers to primal chaos. This primal chaos had been divided by certain properties of light on the second day but the divided reality is still chaos through and through. On Day Three this chaos is for the first time driven back (remember that all this is done as a matter of mathematical operations; nothing material yet exists). The thing that drives back chaos is the creation of the first true number (three) and the corresponding generation of the most fundamental geometrical figure, the triangle.

The text immediately leaps ahead to the creation of something called "Earth." It can make this leap because the "Earth" spoken of in Genesis is not a planet or soil or anything else of a material nature but is, rather, the element "Earth" whose geometrical figure, the cube, was deemed to be made entirely by arrangement of triangles. The text is telling us that once the triangle exists, the various Platonic solids come into existence beginning with the element Earth, an idea that is preserved in Pythagorean teaching.

It is important to underscore that when we speak about the element Earth as known in Greek physics we are actually referring to a three-dimensional geometrical figure known as a cube. This cube is important in the flow of Genesis because it introduces height and depth to a universe that up until this point consisted of nothing but a two-dimensional line. In other words, with the establishment of the number three and the figure of a triangle upon which the cube was built, the mathematical and geometrical principles are in place within the Genesis text for the delineation of a three-dimensional space within the waters of chaos. The mathematical delineation of a three-dimensional space is, in the text, instantly followed by the appearance of life as an extension of the same mathematics.

A study of the Hebrew terms used in Genesis helps to clarify the message of the text. In Hebrew thought, the word rendered as Earth is *eretz,* a term that describes a habitable space. The key word is *habitable.* The prophet Isaiah expressed the idea this way: "For thus says the Lord, who created the heavens, who is God, who formed the earth and made it, who has established it, *who did not create it in vain, who formed it to be inhabited*: I am the Lord, and there is no other" (Isa. 45:18, emphasis added).

The Hebrew term *eretz* does not identify a planet or any other physical object, it expresses a quality. In Hebrew thought, the opposite of earth is any realm in which life is not possible, or in other words, the realm of the dead. In the Bible that realm is, as we might have expected, associated with the term *water* which is, in turn, connected with the idea we know as chaos, a condition in which life cannot survive. The book of Job provides an example of the way the Bible associates water with the land of the dead: "The dead tremble, those under the waters and those inhabiting them. Sheol [the realm of the dead] is naked before him" (Job 26:5–6a).

Earthly Existence

The event associated with Day Three in Genesis is the writing of the mathematics that generate a field within which life can flourish. That field is the precondition for a three-dimensional habitable universe. It is important to remember that in Genesis the thing that has been made is an informational field and not a material environment. Genesis is describing the writing of a blueprint for a system of communication rather than the construction of a series of objects. For Genesis, all that was necessary for the emergence of life was the invention of the number three and with it the coding of the mathematics and geometry supporting three-dimensional space. As a careful reader can see, once a third dimension is in place, the text of Genesis is suddenly bursting with life. Genesis unfolds this way because in a universe defined by information, once a third dimension has been coded into the mathematics of reality the foundation has been laid for the existence of life. All that is necessary is the arrival of a human being whose observations and measurements will reduce the mathematics to physical form.

This idea of an inevitable flow between the number three, the creation of the triangle, and the arrangement of a three-dimensional cube-shaped geometrical structure called the element Earth, all of which provides the mathematical condition necessary for life, may seem complex but the semantics of Genesis make it easy to understand. In Genesis 1:11 we read, "And God said, 'Let the earth bring forth grass.'"

Any poll conducted among believers or nonbelievers that posed the question, "Who, in the Bible, created life?" would surely yield an almost unanimous answer, "In the Bible God created life!" Surprisingly, that answer is not exactly right. Genesis is quite clear, it is the element Earth that "brings forth" life. At first blush this may sound like an affirmation of some of the claims of evolutionists. That hope dissolves, however, when we recall that Genesis is focused on the coding of numerical and geometrical information. The Earth spoken of here is neither a planet nor is it soil, it is a three-dimensional geometrical principle expressed by the figure of a cube.

We can summarize the story of the origin of life as seen by both Pythagorean thought and its precursor in Genesis in simple terms. With the creation (in principle) of the number three and with it the triangle, the precondition for the organization of the platonic solids has been established and the foundation for a three-dimensional habitable universe has been laid in the midst of the seas of chaos. Once a habitable universe is in place, it will certainly be inhabited.

Perhaps the most astonishing aspect of Genesis is the way in which it invites us to think of life as a byproduct of certain principles of geometry. Every time I read Genesis, I am astonished anew at the casual attitude the text takes toward the origin of life. I suspect that there is a clue here that can help us unravel some of the mysteries that fog that issue in contemporary science. Scientists have a tendency to think in terms of extraordinary events and difficult arrangements when contemplating the beginning of life. Their experimental attempts to replicate the origin of life lean toward the exotic, blending odd atmospheric gasses with almost unthinkable bursts of energy. Genesis is telling us that all you need, really, is a three-dimensional space and life will reliably appear.

There is at this point a surprising corollary between the informational approach of Genesis and a recent discovery by materialist science. Materialist analysis tells us that the synthesis of protein, which is the process most basic to life, involves interactions in three dimensions. That is, the information encoded on DNA is transcribed to messenger RNA in three dimensions, not surface to surface. The three-dimensional aspect of transcription seems crucial to the process. The chemicals involved must bond in three dimensions or they will not communicate the necessary information. This means that in both Genesis and current materialist science, the prerequisite for life is a version of reality that possesses a three-dimensional structure. In such a system no God ever has to say, "Let there be life." It is enough to establish three-dimensional reality and reality itself will bring forth life. This insight may help clear up some current questions in science and religion.

Ubiquitous Life

One of those questions has to do with the ubiquity of life in the universe. Darwinian evolution would be gravely harmed by the discovery that life has independently arisen on another world like Mars or Europa. The chance that materialist processes could accidentally generate life are ridiculously small, even when we restrict our scope to one planet. When the claims of evolutionists are subjected to statistical analysis, the chance that the conjectured sequences of accidental events that they propose occurred at all falls in a category of probability technically defined by physicists as "absurd." Should life appear in another place, the materialist approach to origins would have to be entirely abandoned because materialist science would have to prove not only that a sequence of absurdly unlikely events occurred but that they happened time and again. No theory could bear that weight of proof. This is why some evolutionists like Francis Crick and Richard Dawkins are open to the wild speculation that space aliens may have seeded life on earth. They have to offer conjectures of that sort to protect themselves against the possibility that life will actually be found elsewhere. However, if Genesis is right, and life is the inevitable byproduct of mathematical and geometric information that is ubiquitous throughout any three-dimensional universe, then we should not be surprised to find life almost everywhere we look. It will be interesting to watch this issue play itself out as we learn more about what really hides in the soils and water of our solar system.

There are some other points that are clarified by the Day Three text of Genesis. One of the keys to reading and understanding Genesis is an appreciation for the way in which Genesis and the Bible as a whole use specific examples that are provided in the text. In the Day Three text, Genesis speaks specifically about grasses and fruit bearing trees that seem to spring to life. This has occasioned some confusion as creationists try to reduce all this language to some sort of neomaterialist scenario that finds them contending that the first thing that appeared on earth was grass followed by fruit trees, all of these somehow busting into full bloom before the sun even begins to shine!

These people fail to understand that the Bible takes a consistent and unique approach to the use of specific examples. When a general proposition is offered in the Bible, it is often immediately linked to specific objects, persons or events. These examples are not meant to be exhaustive but illustrative. In Genesis 1 the references to grass and fruit trees are intended to clarify what the text is talking about when it talks about life. The same is true when a few creatures are singled out for mention on other days in Genesis. These are all heuristic examples intended to leave no doubt as to the meaning of a particular passage.

The reference to common things like fruit trees and grasses on Day Three serves several functions. First, it tells the reader that when the element Earth and its expression as a three-dimensional Platonic solid establish the mathematical preconditions for life, the life in question is real life just as the reader knows it, the sort of life the reader could encounter any day of the week. This protects the text of Genesis from being read as some sort of philosophical musing by connecting its physics directly to the world we actually experience every day. Second, biblical examples define the range of entities under discussion. In this case, grass and fruit trees are very different things. In the Bible, grass is paradigmatic for things that are here today and gone tomorrow but fruit trees are treated as something of lasting value that is to be preserved and protected. The point that Genesis is making by using these two examples is that the full range of life finds its origin in the three-dimensional geometry of the cubic element called Earth. At the same time, it is significant that no animals or birds or fish are mentioned at this point. This does not mean that these other creatures are born of different forces; they are not. The text of Genesis will go out of its way to make sure that we understand that these entities are "brought forth" by the element "Earth." But the Spirit that inspired Genesis will make the point that these other beings also reflect the operation of elements other than Earth, elements like "air" and "water." The effort of some to turn Genesis into a step-by-step natural history of life on earth will generate a muddled hash and will prove unfaithful to the intention of the text. We need to focus again and again on the fact that what is being described in Genesis is the writing of a digital code, a software package that provides the content behind the appearances ("seemings" in Greek) that we know as our material universe.

The reference to trees and grasses also allows the text to introduce the concept of reproduction where, according to the text, each kind of being brings forth young "according to its kind whose seed is in itself" (verse 11 repeated in verse 12). It is interesting to note that this remarkable sentence actually introduced the concept of "species" to the world. The real Origin of Species, at least in terms of the concept of a class of creatures able to reproduce themselves according to their kind, is, historically speaking, first encountered in Genesis. But even more important, this reference to bearing "according to its kind whose seed is in itself" established a very important principle, something I call *derived naturalism*.

Closed System, Open Book

Naturalistic materialism is the idea that the universe constitutes a closed system in which all inputs are generated within the system itself. We have seen that the materialist idea is a pipe dream for the simple reason that life consists primarily of information and no chemical or physical material can generate information (ink and paper cannot write a book). There is, however, another error that can be made in the opposite direction. A person might credit everything that happens to an outside agency, such as God. That sort of thinking denies to the creation any integrity or meaning. God created a habitable universe for a reason. That reason can only be served if the forces coded into that universe are allowed to play themselves out with some independence from God.

That independence is affirmed in Genesis by the concept of reproduction of each species after its own kind. What this phrase means is that the fate of each species (including ourselves) is at least partly in our own hands. The creatures of Day Three are not automatons run by some demiurge who controls their every movement. They are beings who are responsible for maintaining their existence generation to generation, each according to its own kind. Presumably, this means that creatures (including people and nations) are free to live their way into extinction. The point of the text is that creatures and communities must write their own history within the realm of materiality. The picture Genesis gives at this point hints at Deism without actually going there. Deism is the opinion that God is a watchmaker who builds the watch that is the universe, winds it up, and lets it run. In some measure, this is an accurate picture. God may know about the fall of every sparrow just as Jesus said, but that does not mean God determines every movement by every sparrow. Such a vision of God would quickly fall into pantheism, a doctrine in which God and nature end up being the same thing.

Deism will never be an option for biblical faith. The biblical testimony to God's great love for Israel and His dramatic intervention in history in which He sends His own son to die for us all puts the lie to any idea of a watchmaker God. However, it is also true to say that neither does the God we meet in the Bible meddle in every little event of life. Even in the Bible God is an elusive presence. He does not continually dominate the story that is told there. He enters the story at key points but then seems to vanish for long periods of time. God gives humanity and presumably the rest of creation the freedom to do stupid things. Not every crisis is a test sent by God; sometimes we make our own problems. Not every moment of good fortune is, biblically speaking, a blessing from God; sometimes we make our own luck (although usually we would have been better off waiting patiently for God's blessing). The main field of God's activity in the Bible is pretty tightly focused. It centers around Jesus Christ, in the prophecies, events, and rituals that point forward to Him from out of the past and in the history of two communities (the church and Israel) that each find their destiny in His future. The Bible has surprisingly little to say about many aspects of world history. Wars come and go without a divine comment of any kind. Some kinds of technological development are of no interest to prophecy (I have yet to see a biblical reference to the invention of the automobile). There are a host of matters left out of the Bible simply because these things are not part of the main story arc of the history of God with the world.

There is, then, a kind of *derived naturalism* at play in the daily life of the world. The main story of creation and redemption is in God's hands and that story is, unquestionably, carefully directed by its author. But there are fields where naturalistic materialism actually has it right and the processes of nature work themselves out quite well without divine interference. The Genesis Day Three account tells us this when it announces that the code is now in place for a world in which the creatures of the earth will be able to live their lives and reproduce themselves each in their own seed and according to their own kind, generation after generation without further divine interference or command. It is a picture in perfect accord with materialism except that this freedom is a freedom that has been bestowed. It is, in other words, a freedom that has been *derived* from God rather than carved out in opposition to God like the freedom claimed for creation by materialism's high priesthood.

Faithful Living

This manner in which Genesis speaks of God's rule can help resolve one of the most vexing problems in the history of Christianity, the problem of God's sovereignty vs. human free will. This problem, which has divided denominations and split fellowships for centuries, can be stated in this way: If God is Creator and Lord and if God possesses governing power over every molecule of creation, what space can there be for free will? But if there is no space for free will, how can we be held accountable for our choices as the Scripture clearly says we are? This question becomes especially urgent when applied to the issue of salvation, where it can be stated in this way: Are we saved by God's choice (of us) or our choice (of God)? Suffice it to say that biblical texts can be marshaled to support either side in this debate. For our purposes here, I want to indicate the contribution that Genesis 1 can make.

In Genesis, we are confronted by a world that is the product of numerical codes and archetypal patterns. Everything in our universe is understood to be the expression of these simple patterns. This gives to the universe a God-directed, predetermined quality. In a sense, the life of the universe really does consist in the reiteration of predetermined stereotypical sequences within which there is little space for free will.

This tightly determined universe is, however, the universe as seen from a *heavenly* perspective. That is not the perspective from which we normally live our lives. In our world, as far as we can see, each species sustains itself "according to its own kind." In other words, we normally see no divine predetermination evidenced in the flow of daily existence. This split between the universe as it is (the heavenly perspective) and the universe as we experience it can resolve the ancient problem of God's sovereignty and man's free will.

Day to day we live in a bubble universe that appears to be the very embodiment of the closed universe of materialism. But when we look more closely, the bubble appears to have been cunningly designed, a fact even Richard Dawkins, Francis Crick, and other materialists freely admit. Biblical revelation sketches this design by sharing technical information about the blueprint upon which everything is built. It also tells stories and narrates history with an eye toward pointing out repeating patterns. But revelation carefully restricts its content to what God wants us to know and does not go beyond that.

The challenge set before us is to live our lives in the universe that we know while letting our decisions be guided by the revelation that we have been given. This way of living might best be described as obedience. That is, we turn to revelation and study the Bible to understand what has been shown to us, but then we turn back to the world and, informed by revelation, make the best choices we can in the context of the issues set before us each day.

On the matter of salvation the issue is not one of this way or that, it is a matter of perspective. Are we saved by God's choice of us? The answer is, emphatically, yes, from the heavenly perspective. Are we saved by our choice of God? Again, the answer is just as emphatically yes, from our perspective within the bubble-verse of creation.

This way of thinking preserves the distinction between heaven and earth and that is an important thing to do. The greatest problems in daily living result either from trying to import the things of earth into the realm of heaven (giving earthly issues an importance they do not deserve) or from trying to force upon earthly life the authority of heaven (thereby denying to people the right to make their own mistakes). Creationism, for example, uses fossil and rock sedimentation evidence, all of which is nothing but the temporary and contingent product of human observations and measurements, to make truth claims about eternity. In doing this, creationism evolves into blasphemy. On the other side of the issue, the doctrine of predestination as presented in hard-line Calvinism tries to assert the millennial authority of Christ over every detail of life before His glorious appearance. Such an assertion robs material life of its God-given dignity and strips human history of its significance.

Faithful people are called to live by faith and living by faith means living in the shadowlands between hope and sight. We have the truth vouchsafed in the Word of God but that truth is not yet the truth of our world. We experience the authority of God's rule proleptically in the church or in the history of Israel, but that authority does not yet encompass nature or even the history of nations. As a result, people of faith live in the world but we are not of the world, an oft-repeated cliché that nevertheless translates into a very real and challenging way of life. We endure the burden of living each day as if everything depended on us doing our job because we are commanded to live that way, while at the same time we remember that everything depends on God doing His job because that is the truth that gives our efforts meaning.

All of this is part of what I call *derived naturalism*. It means that on a day-to-day level, lives of faith will look pretty much like all other lives. We will make choices in a world that Richard Dawkins or Francis Crick would fully recognize. But people of faith will remember something these men consciously choose to forget. People of faith will remember that everything that we do, and, in fact, everything that exists, derives from a wisdom, logic, and eternal pattern that, while not fully understood by us, nevertheless determines the fate of the universe. People of faith, aware of aspects of this greater pattern, will strive to conform themselves to it as best they can and in the process will produce lives expressive of a balance and wisdom that is not entirely of this earth.

Chapter 7

A Time for Every Purpose under Heaven: Day Four

Day Four presents the most interesting problems for Bible scholars. Because of this, it also offers the greatest insights for anyone willing to work seriously with the actual text. My own efforts to understand Genesis have always begun with the Day Four enigma. For a long time I had thought it possible to connect the claims of materialist science with Genesis in a linear way. If you look at Genesis superficially, you can see a kind of evolutionary outline in chapter one. The text begins with something primal and big bang–like, the creation of light, and ends handily enough with the appearance of homo sapiens. It all looks sort of scientific in the old-line materialist sense of that term, as long as one does not look too closely.

The problem is that while the devil may from time to time lurk in the details, truth often makes itself known in the odd little things that don't quite fit our expectations. Think back to the geocentric view of the universe where the system worked so well. It could, for example, predict eclipses with complete accuracy; the only problems involved tiny perturbations in the movement of Mars. Yet when those minor errors were scrutinized, the theory turned out to be totally wrong. Any materialist reading of Genesis is similarly undone by the content of Day Four. Let's look at the verses:

> Then God said, "Let there be lights in the firmament of the heavens to divide the day from the night; and let them be for signs and seasons, and for days and years; and let them be for lights in the firmament of the heavens to give light on the earth," and it was so. Then God made two great lights; the greater light to rule the day, and the lesser light to rule the night. He made the stars also. God set them in the firmament of the heavens to give light on the earth, and to rule over the day and over the night, and to divide the light from the darkness. And God saw that it was good. So the evening and the morning were the fourth day.
>
> —Gen. 1:14–19

After a long, fruitless struggle to interpret these verses out of existence, I finally realized that there were only two choices possible here. Read in materialist terms, Genesis is saying that the earth and even complex life forms like grass and fruit trees (for heaven's sake!) burst into existence on Day Three before the sun began to shine. This seemed to me obvious nonsense. But that did not mean that the text itself was nonsensical. The fact that the text did not make sense when read in a materialist way sent me to exploring the approach I am following in this book. Since I began my work in that new direction, I have begun to appreciate Day Four of Genesis as a gift. Far from being an embarrassment that needs to be quickly passed over with some mumbling excuses and vague comments (the usual approach in evangelical scholarship), Day Four opens the door to the worldview of Genesis and to revelations about our universe that otherwise might have remained hidden behind that door. Day Four helps us to grasp the opportunity Genesis offers to see our universe not through the narrow aperture and blurry glass man uses but from the clear and broad perspective of the Creator.

Day Four is not a tale about the construction of sun, moon, and stars. The order of the days confirms the illegitimacy of that reading of the text. Genesis 1 tells us that on the fourth day of creation God established the principle of Four. This arrangement of the days supports my contention that Genesis is not the history of materiality; it is, rather, a description of the software program that defines our universe. In Genesis each new layer of the digital code underlying the operation of our universe is unveiled day by day for our inspection. The eventual form this program will take as a physical reality is not the focus of the text. For Genesis, once the program is set in place, the words of Ecclesiastes govern what happens next: for everything there will be a season.

So far in Genesis we have seen the origin of three dimensions. These have been presented in geometrical terms as the creation of the singularity (the point), the line (two dimensions), and the triangle (three dimensions including height and depth by virtue of the expansion of the triangle into the Platonic solids). Now we encounter a fourth dimension: the dimension of time.

Is Genesis really describing the creation of the fourth dimension in these verses? Many of us were taught that Day Four told the story of the formation of sun, moon, and stars, but the text of Genesis goes out of its way to make its purpose clear. As we have seen, throughout Genesis there are references to various sorts of material objects and creatures (fruit trees, grass, birds, etc.). These specific references are offered as helps to the reader who knows only the objects of material existence. They tell the reader that these are the sorts of things the text is talking about in digital and geometric terms, rendered now into concrete examples so s/he will understand. As heuristic examples they serve as signs identifying the topic under consideration.

In the Day Four material, the text references sun, moon, and stars but it clearly does so in terms of their function in the act of measuring time. These things are mentioned not in their context as material entities but in the role they play in "dividing the day from the night" and serving as "signs and seasons, and for days and years." Genesis is signaling its readers that the topic of Day Four is not the construction of objects but the origin of the thing people measure by sun, moon, and stars. The topic at hand is the origin of the fourth dimension, the dimension of time. Next to that event, the actual emergence of stars is a secondary consequence, the result of human observation and measurement. Genesis is, as always, the story of the composition of information.

The Mystery of Time

The fourth dimension is one of the extraordinary mysteries of our universe. For most of human history we thought of time in terms of a ticking clock. It seemed absolute and predictable. Most important, under conditions of observation and measurement it seemed a thing quite different from the other three dimensions.

Since Einstein, we have come to understand time as a physical dimension, just like the other three. I pointed out earlier in this book that time responds to gravity, mass, acceleration, and velocity. This means that time exists as a form of space; it is not different from other space. We encounter it differently, there is no doubt about that. But it seems certain at this point that the reason for the difference between our encounter with time and our encounter with the other three spatial dimensions is rooted in the limitations of our experience and not in the nature of time. When we measure time we are measuring a spatial dimension that we encounter in an odd way.

We have also come to understand that time is protean and not at all fixed. The story of twin brothers has often been told but people do not always appreciate how disturbing it really is. Imagine two brothers, born the same day in the same hour. If one brother journeyed at half the speed of light to the nearest star, he would, upon return, be a bit more than two years younger than his sibling. Think about that a moment. The traveling brother would not just look or feel younger, he would in fact be a younger man. The twins would not have the same birthday anymore. Time is not fixed. Its flow is governed by forces like velocity that have the capacity to dramatically change the answer to the question, "What time is it?"

One way to illustrate the protean nature of time is to take a fresh look at some scientific questions in dispute today. How old is the universe? The answer depends on how fast you are moving or how strong the gravitational field is that surrounds you. It turns out that any answer will be correct depending on the circumstances of the person taking measurements. Is the universe five thousand years old (in earth years)? Absolutely, at some speed or in some gravitational field. Is it five thousand trillion years old (again in earth years)? Certainly. Send two vehicles hurtling by each other and there is no telling what sort of time relationship might form within the vortex of their relative motion. Time knows no Archimedean vantage point. Science fiction writers love to speculate about "time distortions." But their language makes no sense. How can there be a "distortion" when there is no correct reading against which a "distortion" could ever be measured? Time is one of those things that simply is what it is for you right now. Nobody can ever say you are wrong in your measurement of time. This is what Einstein was telling us when he applied the word *relativity* specifically to time.

I find it very significant that the Bible, inspired by the one who wrote the blueprint for existence, always took the point of view taught by Einstein, a point of view that still, to this day, commands almost zero understanding in the broad public mind. Einstein's theory of relativity provides the context in which we should read the often quoted words of Peter: "But, beloved, do not forget this one thing, that with the Lord one day is as a thousand years, and a thousand years as one day" (2 Peter 3:8).

This verse is sometimes twisted by creationists who try to force from it a hint that maybe the six days mentioned in Genesis might conceal vast geological ages. We have seen that cannot be the case because other references throughout the Bible all insist that the universe was created in six standard earth days. The creationist argument misses the real point of Peter's statement and robs it of its startling revelation. What Peter is telling us is that the flow of time depends on who is doing the measuring and the conditions in force at their location. He reminds us that God is not measuring time from a location somewhere in Jerusalem or anywhere else on earth. In that context his reference to the "number" one thousand is telling. The term *one thousand* is one of those symbols that conveys more than it seems. In the ancient world, the term sometimes served as a stand-in for Carl Sagan's proverbial "billions and billions" of years. It suggested a quantity beyond measurement. Under the guidance of the Holy Spirit, the craftsperson who wrote the code of creation, Peter is describing the flow of time as a quality that, in different locations and under different conditions, can vary by a factor that might require the ancient version of "billions and billions" to express the difference. This is nothing less than Einstein's theory of relativity outlined two thousand years before his time.

But St. Peter was not the first to observe the relativity of time. In the Old Testament we encounter these startling words: "For a thousand years in your sight are like yesterday when it is past, and like a watch in the night" (Ps. 90:4).

Once again we see the reference to the protean concept of "a thousand years," a concept that grants to time extraordinary elasticity. What makes these words startling is the fact that Psalm 90 is credited to Moses, the same man who gave us the book of Genesis, including chapter one, and who by claims of the ancients provided the inspiration for Pythagorean and Platonic philosophy. Moses himself understood that time is flexible and subject to various forces.

There is another area in which the Holy Spirit anticipated the discoveries of science by quite a margin. Genesis 1 tells us that time had a beginning. This insight always contradicted the prevailing view that held sway over most of human history. Prior to the twentieth century, time was simply accepted as a given part of existence. It was not only absolute, it was also thought to be eternal. Science no longer accepts this assumption. Once it was clear that time is a physical dimension, it also became clear that time resembles the other three dimensions we encounter in this universe in that it must have had a beginning just as they did and will most certainly have an ending, exactly as the Bible always told us.

We also understand today that just as there is a minimum unit of space beyond which we cannot measure (the Planck length) there is also a minimum unit of measurable time (10^{-43} seconds), which is the time needed for light to traverse the Planck length. This means that there are energy levels (quanta) in which time does not exist as far as we can ever know.

The upshot of all this is that time and space exist within a narrow frame. Genesis describes the creation of this frame. The book of Revelation at the end of the Bible describes its dissolution.

The Day Four Genesis text accurately limits the realm where time holds sway to the narrow bubble-verse we measure and observe into material existence. This means that when humanity dies, time dies.

The Mortality of Time

The recognition of the mortality of time is important because it reaffirms what we are learning from the Genesis text, namely that the universe we know has a bubble-like character with clear boundaries and narrow limits. Since our bubble-verse includes only four dimensions and quantum mathematics tells us that there seems to be at least ten, there must be a realm completely outside time and space. This means that the time-bound universe we experience exists only in a temporary and contingent way and that time is not a universal or fixed part of reality per se. Einstein spoke of this when he said, "People like us who believe in physics know that the distinction between past, present, and future is only a stubbornly persistent illusion."[1] This idea that time is relevant only within the boundaries of the reality we have observed into existence is amplified in quantum physics.

Quantum theory presents an extraordinary new concept of time. This approach is well expressed by physicist Michio Kaku, who in describing the thinking of Andre Linde, one of the founders of the inflationary theory of the universe, offers this remarkable summation: "According to Linde's philosophy, dinosaur fossils don't really exist until you look at them. But when you do look at them, they spring into existence as if they had existed millions of years ago."[2]

Kaku is referring here to the fact that when we collapse the wave structure that is the normal state of the universe and temporarily reduce reality to its material expression for purposes of analysis and measurement, the history of anything we see is a byproduct of our observation and not a quality possessed by the object that is being observed. In other words, we do not see a fossil that is fifteen million years old; we see a fifteen-million-year-old fossil. There is a big difference between the two claims.

In quantum time theory we cannot speak of an objective time line moving inexorably from past to future independently of an observer. In quantum physics nothing happens independently of an observer. Instead, we must understand that each past and future is a subcategory of a particular present temporarily generated by someone's observation. Each observation generates its own past and future without reference to anything that has gone before or that might come after. The question then becomes, why do our individual observations tend to produce pasts and futures that seem to be consistent with each other? Physicists have no answer to this but assume that the mathematics of time must in some way make one particular commonsense flow of events the most likely state that will emerge from each isolated act of observation. In other words, our individual observations tend to generate a material universe that falls consistently into the same rut.

Quantum theory insists, however, that there is no necessity that this should occur; quite the contrary. Quantum physics holds that everything is possible as a matter of principle and that anything that is possible in principle must eventually occur in fact. This insistence is necessitated by the law of the conservation of energy which holds that energy can neither be created nor destroyed. Since quantum potentiality exists in the form of energy states, every element of that potentiality must find expression in some way because to propose that energy can just disappear is scientific heresy.

Taken as a whole, the theory of time that results from all this has some strange and disturbing implications. For one example, it may happen that when we collapse the wave structure of infinite possibilities that Abraham Lincoln consistently emerges as a past president of the United States. There is, however, no reason to exclude the possibility that as part of someone's observational activity Stephen Douglas might at any moment emerge as a past president in his place. The truly disturbing fact is that not one of us would ever notice or contest the change. At any moment any of us could, in principle, find ourselves part of a new family, a family we would protect and care for with all our love because they would be the only family we had ever known. It is, in fact, quite certain in light of the standard form of quantum theory that such a thing will happen at some time, we just don't know when or where such a leap will occur. Every time we observe the world, the present, past, and future that result are all a matter of pure chance.

It was in response to this aspect of quantum theory that Einstein supposedly objected that "God does not play dice with the world!" People remember that complaint but forget Neils Bohr's apt reply, "Albert, stop telling God what he can or cannot do!"

It is, actually, at this point that God can come to the rescue of quantum physics itself. As I said, quantum physics has no way to account for the continuity we experience in our lives and the history of our world. This has provided ammunition for criticism of the theory. The Bible, however, addressed this problem long before Neils Bohr and company first drew breath. In his letter to the Colossians, Paul said this about Jesus Christ:

> By Him all things were created that are in heaven and that are on earth, visible and invisible, whether thrones or dominions or principalities or powers. All things were created through Him and for Him. And He is before all things, and in Him all things *consist.*
>
> —Col. 1:16–17, emphasis added

The emphasized word in that passage renders the idea of holding things together. Paul is telling us that it is in Jesus Christ that everything coheres or is brought together in a way that makes common sense. This claim is repeated in the Letter to the Hebrews:

> God, who in various times and in various ways spoke in time past to the fathers by the prophets, has in these last days spoken to us by His Son, whom he has appointed heir of all things, through whom also He made the worlds; who being the brightness of His glory and the express image of His person, and *upholding all things* by the word of His power, when He had by Himself purged our sins, sat down at the right hand of the Majesty on high.
>
> —Heb. 1:1–3, emphasis added

Once more, the text tells us that Jesus Christ holds everything together with the implication that should he cease to do so our world would dissolve into chaos in an instant. Quantum physics works hard to square its claim that everything is possible with the reality that everything that could happen does not actually seem to happen. It has no technical explanation for the apparent continuity of existence that seems to violate the law of the conservation of energy. This is the point where the Bible offers a scientific insight that can prove useful for quantum theory. It may be that everything that could happen does not happen because there is a restrainer who holds things together. The presence of such a restrainer would simplify and streamline quantum physics. While all things would remain possible, they would remain possible only for the restrainer. The realm of space/time would be limited to the parameters the restrainer imposed. Unexpressed possibilities would achieve full expression in His life in a way that is consistent with the law of the conservation of energy, but they need not happen to us. The Bible itself actually makes precisely this claim when it says: "For it pleased the Father that in Him all the *fullness* should dwell and by Him to reconcile all things to Himself, by Him, whether things on earth or things in heaven, having made peace through the blood of His cross" (Col. 1:19–20, emphasis added).

The word translated *fullness* in this passage is *pleroma*, a technical term in Greek philosophy that expresses the totality of potentiality. The verse contends that all possibilities do in fact come to expression but not necessarily in the bubble-verse we inhabit but rather in the limitless life of Jesus Christ. This point is underscored when the text includes "all things" and "things on earth and in heaven" within the pleroma that dwells in Christ. These terms necessarily include those possibilities that did not and will not come to physical expression as a result of human observation and measurement. This language offers a way for quantum physicists to resolve the problem posed by unlimited possibilities without turning history into a game of chance. At the same time, quantum physics provides a fresh way for the rest of us to think about the Lordship of Jesus Christ over time and history. I find it ironic that the name of Jesus upsets secularists in a way no other name does. And yet that same Jesus may be the very person who, by His grace, is holding the things they value—their family, nation, environment, and world—together moment to moment, allowing them to live meaningful lives in a universe that otherwise perches on the edge of chaos.

It may be that one way Jesus Christ controls the flow of history is by influencing the way in which human beings collapse the wave structure of potentiality and generate the material world. Jesus is, after all, the Word of God; His life flows through the pages of the Bible. That Bible has structured the thinking of much of the world in such a way as to make certain kinds of "Genesis by Observership" (John Wheeler's term) more likely than others. This limitation applies to both believers and nonbelievers because the influence of the Bible on Western culture has been all-pervasive. If this speculation is accurate, the current effort to strip our culture of biblical influences, especially our schools and government, might be a very bad idea. We may be opening a door to chaos beyond our imagining. It may also be important that the Bible speaks often of God's Holy Spirit coming to dwell in the minds and hearts of believers. We have seen that the Holy Spirit embodies God's original blueprint for creation. The presence of Spirit-inspired believers among those collapsing the unseen quanta into physical existence may also be a restraining factor that keeps material reality from flying off in all directions. In light of that possibility, it is interesting to note that in the New Testament the dissolution of the cosmos begins with the rapture or removal of the church from history.

The idea that the realm of time is limited to the scope of our experience helps us understand the Bible in one more way because the Bible wants us to understand that it was written from a perspective that is specifically identified as outside the time domain. The Bible makes this claim again and again throughout its pages by pointing to the accuracy of the prophecies it contains. We saw one example of this predictive capacity in our study of the book of Daniel. The idea is that such extraordinary accuracy is possible only because there is a vantage point from which time can be seen in all its possibilities and from beginning to end, like a parade viewed first band to last float from high above the parade route.

The notion that there is a realm beyond time is important because it implies that there is a limit to relativity. We have discovered that there is no Archimedean point within time. There is no location from which we can say that this is the real "now" or that history as we see it in this moment is the past as it really unfolded. From this it clearly follows that no person who lives within time/space can ever claim to offer more than a wild guess about the best course to follow in light of the future prospects of our world. But if there really is a realm beyond the limitations imposed by time and space, then cosmological and moral teachings that originate in that realm can demand a higher level of respect because they are not attenuated by the relativistic murk of the bubble-verse.

Statements that find their origin outside time and space present themselves to us as revelations, truths we could never discern on our own. These are exactly the sorts of statements the Bible claims to offer. The fact that some of those statements may contradict what we perceive within the bubble-verse does not obviate their truth in any way. The usefulness of revelation must be measured by its capacity to reveal what we do not know, not by its capacity to confirm what we think we know. But how do we know if something is a revelation as opposed to an illusion spun by someone as lost in the seas of relativity as we are? The Bible tells us that there is only one way to separate truth from illusion and that is to test the capacity of a text or a teacher to accurately describe the future before it happens. The ability to do that is the thing that confirms whether or not a message is really coming from outside the bubble-verse, from the one who gives life its continuity.

The Bible tells us many times that it is its ability to accurately predict the future that assures us that the moral claims and technical information about our world that it offers are right. God himself calls us to respect His Word by offering this reminder: "Remember the former things of old, for I am God, and there is no other; I am God, and there is none like Me, *declaring the end from the beginning, and from ancient times things that are not yet done* (Isa. 46:9–10a, emphasis added).

God sets out the test by which a teaching (or prophecy, which means the same thing) is to be measured: "And if you say in your heart, 'How shall we know the word which the Lord has not spoken?' When a prophet speaks in the name of the Lord, *if the thing does not happen or come to pass, that is the thing which the Lord has not spoken;* the prophet has spoken it presumptuously; you shall not be afraid of him" (Deut. 18:21–22, emphasis added).

Jesus' life fulfilled to the letter more than three hundred prophecies that, centuries before His birth, spoke of a coming savior. That fact alone tells us that we ought to pay attention to the message of Genesis, not because Genesis is a religious book or is in some way "holy," but because that kind of precision in predicting the future suggests that the entire Bible including Genesis might well provide a look at our universe from beyond its boundaries, an authoritative perspective we could never gain on our own.

One+Two+Three+Four

Throughout this book we have been turning to Pythagorean thought to gain clues into the message of Genesis. This is based on the ancient claim, often repeated, that Pythagoras drew his wisdom from studying the books of Moses. The main insight that emerged from the Pythagorean tradition was well defined by one of the scientists that gave modern physics its shape: "From the intrinsic evidence of his creation, the Great Architect of the Universe now begins to appear as a pure mathematician."[3] The astronomer Galileo made the same point in a comment about the value of the Pythagorean approach for the way he conducted his research in the field of natural history: "I know perfectly well that the Pythagoreans had the highest esteem for the science of number and that Plato himself admired the human intellect and believed that it participates in divinity solely because it is able to understand the nature of numbers. And I myself am inclined to make the same judgment."[4]

For Pythagoreans and those influenced over the centuries by their tradition, the thing that appears to us as a material universe is understood to be the visible expression of a digital code. It was the passion of Pythagoreans to study and comprehend this code. As students of Genesis, this should be our passion as well for the tradition cited by James Jeans and Galileo finds its origin in Genesis.

In Pythagoras's system, the number four expressed the idea of perfection. This might be so because, as we will see in a moment, the appearance of that number in the text of Genesis had already marked a point of completion in the work of creation. We have also seen that each number in the Pythagorean system was associated with a geometrical figure. One (a concept, not a number, for Pythagoreans) was associated with the point. Two (again a concept, not a number) was expressed by the line. Three (the first true number) was expressed by the triangle (an object of careful study by Pythagoreans because they drew from it the Platonic solids that introduced the idea of a third dimension). The digit four was expressed by a less familiar but more important figure that Pythagoreans called the *tetractus*. The tetractus conveyed in a powerful way the idea of perfection and finality.

I alluded to the notion that with the number four the text of Genesis reaches a state of completion and perfection. Traditional interpretations of Genesis have always held that the chapter one text can be neatly separated into two parts. The first part of chapter one (days one through four) has traditionally been seen as laying out all the fundamentals of our universe. The second part (days five and six in particular) has always been understood to describe the "filling" of a cosmos whose organization had been substantially completed on the earlier days. It is interesting to note that throughout the medieval world it was universally held that the celestial realm, the realm that is alluded to in connection with Day Four in Genesis, was a kingdom both complete and perfect. This association between celestial objects and perfection may have found its origin in the ancient text of Genesis.

For Pythagoreans, the number four expressed an even deeper sort of completeness. Added together, the digits one, two, three, and four produce the number ten. For Pythagoreans the number ten constitued the *tetractus*, a term that can be rendered into

English as "fourness." The tetractus expressed the deepest secret of Pythagorean thought, a mystery so sacred that its revelation to outsiders incurred the penalty of death. The mystical power of the tetractus was the real secret concealed within the all-important figure of the triangle. The triangle was understood by Pythagoreans to express a perfect "fourness" on each side as can be seen in this image:

*

* *

* * *

* * * *

Note that each side of the triangle is made of four dots and that the triangle is filled in with dots totaling 1+2+3+4, forming the tetractus. This figure represented both the power of the number four and the structure of the completed universe. The tetractus also expressed the total of all the musical ratios and, as such, served as the geometrical and digital expression of the relationship between the planets that came to be known as the "music of the spheres." Johannes Kepler actually used this notion of a correspondence between the tetractus, music, and the planetary orbits to devise the first accurate description of our solar system. I strongly suspect but cannot prove that the seven-day structure of Genesis stands as the inspiration behind the musical scale later introduced by Pythagoreans and that Genesis 1 may have once taken the form of song.

The important point is that with the creation of the digit four, the design of the universe as we know it was complete and perfect. We must of course keep in mind that at this point the universe existed only as information in the mind of God. Genesis still has told us nothing about the material adumbration of this code. We understand why this is so. If the universe is, as Genesis says, a mathematical construct like a virtual reality game, is it really all that important to describe the pictures that appear on the screen? Not if we know the code. The visuals of creation can only be the manifestation of a reality fully present in the digital arrangement of the code. The Spirit that gave us Genesis already knew the quantum law of nonseparability and its requirement that nothing, including the flow of time, comes into material existence apart from a human observer actively taking measurements. If the basic structural codes fully exist as potential, that is, as pure digital information (as they do after Day Four), they will automatically spring into material existence simultaneously with the designing of a sentient being capable of making the necessary observations. It is toward that day that the text of Genesis inexorably moves.

Chapter 8

The Darkness Rises: Day Five

We have been reading the book of Genesis through the eyes of the ancients, in particular the Pythagoreans, a group whose intellectual heritage has been traced back to Moses. In his book, *The Introduction to Arithmetic*, philosopher and mathematician Nichomachus described the Pythagorean point of view in these words:

> All things in the world that have been arranged by nature in an artistic process ... appear to have been distinguished and ordered according to number, by providence and the Intellect that creates the universe. For the pattern was fixed, like a preliminary sketch, by the controlling role of number, preexistent in the thought of the world-making god, number intelligible only and altogether immaterial, and yet the true and eternal Being, so that by reference to this as to an artistic plan (*logos technikos*, in Greek) all these things should be produced, time, motion, heaven, stars, revolutions of all sorts.[1]

Nichomachus's description of Pythagorean cosmogony precisely summarizes the process of creation described in the Genesis account. The creation took just six days, as confirmed by various references throughout the Bible. The act of creation reported in

Genesis involved the construction of a "preliminary sketch" in the form of the digital model that we are examining right now. This sketch was "altogether immaterial," which means that it took the form of what Nichomachus called a "logos technikos" or, as we would say today, a blueprint. Since nature is the product of this schematic we can say with Nichomachus that for Genesis all of physical reality is the expression of a fixed "pattern" that has been "ordered according to number."

Today we would probably update Nichomachus's second-century language and describe the Genesis process in terms of writing a software code. On the one hand, Genesis does not describe the construction of any material object. Only numbers and geometric figures are produced in the days described in Genesis. On the other hand, nothing will come into existence that is not present as potential somewhere in the Genesis code. From the biblical point of view, the content of creation depends entirely on the creative efforts of the code writer, and the task set before anyone who would understand how the universe works is to decipher the software code either by temporarily collapsing small segments of the mathematical reality for purposes of measurement and analysis or by attending carefully to the content of the code as revealed in Genesis and the rest of the Bible or, best of all, by doing a bit of both.

The Genesis code that we are describing here looks very simple. But then, codes often are simple. DNA is a digital code defined by the arrangement of just four characters. Most software programs depend on an alteration of zeroes and ones. In the final chapter of this book I will present some adumbrations of the Genesis code that will demonstrate how that code can fully account for the complexity of existence as we experience it.

For the moment we can move on to Day Five of Creation. Once again, the significance of the day is born by the number that defines it: the number five. Throughout Genesis, the creative content of a particular day is defined by its number and the results produced by the day confirm the significance of the number. When we direct our attention to the Day Five events by focusing on the number five, it will become quickly apparent that we are about to gain a surprising insight into both the message of Genesis and the nature of our world.

First, let's read the text as we encounter it in the New King James translation:

> Then God said, "Let the waters abound with an abundance of living creatures, and let birds fly above the earth across the face of the firmament of the heavens." So God created great sea creatures and every living thing that moves, with which the waters abounded, according to their kind, and every winged bird according to its kind. And God saw that it was good. And God blessed them, saying, "Be fruitful and multiply, and fill the waters in the seas, and let birds multiply on the earth." So the evening and the morning were the fifth day.
>
> —Gen. 1:20–23

At first glance, this text seems a bit strange. To this point Genesis has been characterized by a consistent elegance as one by one the fundamentals of geometry have been set in place. Here on Day Five, things seem less neat. Suddenly there are sea creatures, hints about various "living things" that "move" and, incongruously, birds appear! How are we to read this?

Throughout the text of Genesis, the listings of various creatures and entities serve a heuristic function. They clarify precisely what the topic of the day is. These listings are not meant to be taken as descriptions of material entities. They are materialized examples that specify the nonmaterial information being coded into the system that will, in due time, become the universe we know. I see no reason to treat these verses differently.

The first key to understanding the Day Five text is the word *water*. The text clearly connects the activity generated on this day with water. On Day Three, the element Earth "brought forth" grass and fruit trees. We saw on that occasion that those heuristic references indicated a correspondence between the establishment of a third dimension and the appearance of life in such a way as to imply that in any universe where there are three dimensions there will be life. The Day Five reference is similar but not exactly the same. Once again, a basic element in what will later become Greek physics is invested with power to cause creatures to spring forth. In this case, however, it is the "waters" that do that work and these waters are not said to generate life in quite the same way "Earth" does on Day Three. The waters are only said to "abound" with the specified creatures. Nevertheless, the point remains that whatever emerges on Day Five emerges out of the "waters."

We have encountered the term *waters* on days one and two. In each case the term did not directly reference liquid water. The water of Genesis 1 is the realm of chaos, the dark void that existed before the creation of light. It is the primeval soup divided and organized by the firmament on Day Two. We must keep this meaning of the term in mind because it is not permissible to suddenly change the definition of water in the middle of a text in which the meaning of the term has already been defined. The water of Day Five is still the dark void of chaos that we encountered on Day One.

This helps us make sense of the otherwise inexplicable bestiary presented in this text. The Day Five text tells us that the "waters" abounded with "living creatures," "great sea creatures," and, strangely enough, "birds." It may be that this list of animals represents nothing more than somebody's recollection of an idyllic weekend by the sea. But a closer look at the actual Hebrew terms concealed behind the pedestrian English translation reveals more provocative possibilities.

Monstrous Creatures

As we have seen, the creature lists in Genesis provide signs (heuristic clues) that interpret the meaning of the number created on that day. The Day Five list is the most intriguing of all. The first entities with which the "waters" abound are rendered in the New King James translation as "living creatures." The Hebrew term is transliterated as *sherets*. This word refers to a variety of creatures, all of them somewhat unpleasant. Included in the word are references to "creepers," "swarmers," insects in general, and small reptiles. The Genesis list may be redolent of a day at the beach, but it would be a very bad day indeed because the main rendering connected with this word *sherets* suggests something even worse. The term is mostly used to indicate "winged reptiles that walk on all fours." The Day Five list literally tells us that the waters of chaos brought forth monsters. And it only goes downhill from there.

The text tells us that the meaning of Day Five is also illustrated by the emergence of "sea creatures" or "whales" as the King James translation unfortunately has it. The Hebrew term is *tanniyn*. Normally in the King James *tanniyn* is translated not as whales but as "dragons." That's right, there be dragons in Genesis on Day Five! So far, Day Five has given us creepers, swarmers, monsters, and now dragons. It's a tough day.

But this is only the start. The Hebrew word *tanniyn* receives a different rendering in the New King James translation, a rendering also deemed possible by the old King James translators. The word can be rendered as "serpent." This, of course, is, by itself, a nightmarish term. But what sort of nightmare is this? It is interesting to see how *tanniyn* is used elsewhere in the Bible.

We encounter the following usage by the prophet Isaiah: "In that day the Lord with His severe sword, great and strong, will punish Leviathan the fleeing serpent, Leviathan that twisted serpent and he will slay the reptile that is in the sea" (Isa. 27:1). This reference looks forward to the end times. It describes the victory of God over his great enemy. That enemy is here named as the "twisted serpent" but it is the ancient foe we know as Satan. It is significant that in this apocalyptic image the "serpent" is associated with the "sea" (chaos), the same association we are seeing throughout the first chapter of Genesis. This underscores how deeply integrated the biblical message is from book to book and across the centuries of its composition.

In Psalm 91 this ancient serpent appears again: "You shall tread upon the lion and the cobra, the young lion and the serpent you shall trample underfoot" (Ps. 91:13). Once again, the language is prophetic, peering into the end times to describe the fulfillment of a promise that is made in Genesis 3. Israel's victory in its final conflict is in view and, as in Isaiah, that victory is pictured in terms of trampling the ancient serpent, an outcome brought about on

Israel's behalf by the coming of her Messiah. In each of these last two citations, the reference to Satan is clear and the language points to apocalyptic evil.

Passage after passage could be cited but at this point I can say that the signature creatures identifying Day Five range from the nasty to the outright demonic. Day Five in Genesis has given us monsters, dragons, creepers, swarmers, and now, even a hint of brimstone.

But what about the birds? Doesn't Genesis tell us that birds also define the meaning of Day Five? What do birds have to do with the dreadful entities bursting out of the watery chaos on the fifth day of creation?

In our culture we usually look at birds and see something charming, but this is not the biblical attitude. The different mindset is revealed in the fact that the same Hebrew word *'owph* that renders "birds" also renders "winged insects," things like flies, mosquitoes, and other unsavory creatures. It is worth bearing in mind that even the English language preserves a different evaluation of birds. In older times birds were known as fowl.

The birds of the Bible are nothing like the first robin of spring. These are the stuff of Hitchcock horror. Let me point to some examples. In Jesus' parable of the sower, we read that some of the seeds scattered by the sower "fell by the wayside and the birds came and devoured them" (Matt. 13:4). When Jesus explains this parable to His disciples in private, He tells them that the seed is the message of the kingdom and the birds are really "the wicked one [who] comes and snatches away what was sown" (Matt. 13:19). In this parable, birds are a stand-in for Satan himself. In the book of Ecclesiastes it is suggested that birds can serve as spies for an enemy: "Do not curse the king, even in your thought; do not curse the rich, even in your bedroom; for a bird of the air may carry your voice, and a bird in flight may tell the matter" (Ecc. 10:20). Danger and betrayal apparently lurk among the flocks that gather at our outdoor feeders. (I would suspect this applies even more to those fat pigeons that hog all the food).

This theme is expanded in Revelation 18:2, which describes Babylon, Iraq, the hometown of Satan, as a place abounding with birds: "Babylon the great is fallen, is fallen, and has become a dwelling place of demons, a prison for every foul spirit, and a cage for every unclean and hated bird!" It is also worth remembering that Satan himself is named in the Bible as "The Prince of the Power of the Air" (Eph. 2:2), suggesting again an association between ultimate evil and birds.

In the Bible birds are often credited with bringing judgment against people. In Genesis 40:17, a certain baker dreams that three days in the future he would be carrying a basket full of bread on his head to bring to Pharoah, king of Egypt, when suddenly "the birds ate them out of the basket." Three days later, this baker was hung. In Matthew 24:28, the judgment of the wicked is summarized in these terms: "For wherever the carcass is, there the eagles will be gathered together." In this text, birds provide the means by which condemnation is carried out upon nonbelievers. This theme is echoed in Revelation 19:17–18, 21 which describes the judgment rendered against the armies of Armageddon (who somewhat unwisely try to make war against God). In this text, birds finally reveal their true nature as they cruelly execute judgment upon the forces of unsaved humanity: "Then I saw an angel standing in the sun; and he cried with a loud voice, saying to all the birds that fly in the midst of heaven, 'Come and gather together for the supper of the great God,' that you may eat the flesh of kings, the flesh of captains, the flesh of mighty men, the flesh of horses and of those who sit on them, and the flesh of all people, free and slave, both small and great."

Finally, the Bible often combines a remark about the dragon or serpent with a reference to a very strange entity called in some translations "the owl." The translated term *owl* is very odd and has no known avian referent. Some have suggested that it might refer to an extinct creature, but others have seen a more insidious reference. The use of the word suggests that the ancient serpent (Satan) operates in the company of a demonic being possessing bird-like qualities.

Potent Materiality

We have seen that the description of each day of creation is accompanied by illustrative examples that define the information coded into the system on that day. A careful review of the entities listed in the Genesis Day Five text reveals not the cute robins and shiny goldfish suggested by a superficial reading but instead monsters, dragons, and demonic birds. We even glimpse in this text the shadow of Satan himself. Putting it all together, what is the information content of Day Five? What does Day Five code into our universe?

The answer is not hard to find. Since ancient times, the number five has been associated with the occult. The most potent symbol connected with five is the five-pointed star or pentagram. The pentagram is often interpreted as a symbol of evil. That equation is not entirely accurate. But the symbol is rightly *associated* with evil in that it serves in occultism as a potent means of protecting oneself against magical attacks. The pentagram is said to have power to trap a demon who can then be used by a sorcerer to both protect him and to do his bidding.

The faith of Islam has been deeply influenced by esoteric technologies and occult beliefs. It may, therefore, be no accident that five plays a major role in Muslim life. Islam is built around five pillars: the *shahadah* or confession of faith, *salat* or prayer, the *Ramadan* fast, giving *zakat* or alms, and the *hajj* to Mecca. The faith also claims to have been shaped by five lawgiving prophets: Noah, Abraham, Moses, Jesus, and, of course, Muhammad, and is taught in five catechetical divisions.

This information is useful but in the effort to understand the ancient message of Genesis our principle guide has been the writings of the Pythagoreans, writings said to be based on the teachings of Moses in Genesis. In that light, it is interesting to notice two things about the role of the number five in Pythagorean philosophy.

For Pythagoreans, five symbolized the physical nature of earthly life. This was because five is the sum of the first even and odd digits (two and three) which, as we have seen, gave it the name *marriage*. As the first number that mixed the discrete realms of even and odd, five was considered neither even nor odd but it was itself a mystical force.

We have seen that two (a concept not a number) embodied the power of separation. Where once there was monadic unity, now there are two things. We have also seen that separation is the precondition for conscious awareness. To be aware of something is to know that "this" is "not that."

The number three expresses the field in which life becomes a possibility. With three, the fundamental expression of Pythagorean geometry, the triangle, comes into existence. Triangles by extension into the Platonic solids generate the three-dimensional quality characterizing a universe capable of sustaining life (as reflected, for example, in the three-dimensional chemical interactions involved in protein synthesis).

What happens when we put two and three together? What we get is awareness (two) bound within the limits of this bubble universe (three). This sort of awareness can make choices but only within the context defined by the limits imposed by the firmament, the realm of physical reality. It chains itself to the four-dimensional framework of the bubble-verse and it operates as if reality consists only in what is *physically* experienced within those limits. In the words of the apostle James, consciousness of this type "does not descend from above but is earthly, sensual, and demonic" (James 3:15). For a mind working this way, the only truth is to be found in the forms and operations of materiality. It does not understand that material reality is always artificial, temporary, and contingent. This is a pattern of thinking that does not consider the larger possibilities that reside within the mind and purposes of God. In fact it doesn't think about God at all. This kind of awareness is described throughout the Bible as "worldly," a way of thinking that is associated with "flesh."

Instinct in Service to Life

The type of thinking connected with Day Five is not automatically evil. It is more properly designated as *chthonic*. Chthonic means base, or more precisely, instinctual. Chthonic knowledge is the raw instinct of the wild beast. It is associated with a way of living driven by physical passions, most notably fear, greed, and lust. Chthonic thinking describes the mindset of a being who knows nothing but the closed universe of material existence, and as such, is dedicated to nothing more than the accomplishment of its own selfish ambitions and the fulfillment of its elemental hungers.

It is important to understand that the opposite of life lived at the chthonic level is not good as in good vs. evil. The opposite of a chthonic- or instinct-driven life is a life guided by reason and spiritual discernment. This second kind of life has retained contact with nonmaterial reality. It remains connected to the rational world of pure information (the *logos*) that finds its home in the mind of God.

We can now fully identify the archetype coded into the universe on Day Five. On this day God encoded the instinctual drives, hungers, and passions that would eventually find expression within the lives of the sentient beings that will be programmed into the bubble-verse on Day Six. These drives are base but they are not themselves evil. They are primitive but not demonic. We know this because at the end of the very day when these drives are coded into the bubble-verse, the text tells us that "God saw that it was good" (v. 21). These instincts are even blessed with power. God says to the nonsentient creatures that are cited to illustrate the nature of what was encoded on Day Five, "Be fruitful and multiply, and fill the waters in the seas, and let birds multiply on the earth" (v. 22). God has deliberately planted instinctual drives within the creatures of the world in order to preserve life and give it vitality.

Instinct has been given to serve life but sometimes life can become subservient to instinct. When that happens, a line is being crossed. It is, for example, acceptable for hunger to be filled by killing. But when killing itself becomes a hunger, then the being filled with that hunger can be described as evil. It is God's intention that sex should help sustain the life and culture of sentient beings. But when beings or their culture become enslaved to sex, those beings and their culture have slipped into a degraded form of existence.

Genesis Day Five helps us to distinguish what is evil from what is merely instinct by showing us the intended place of instinctual drives in the blueprint of existence. Evil arises when something good has been twisted out of its natural role. Instincts designed to sustain life can take on a life of their own. A primal instinct like fear protects living creatures. But if a creature should start to live in such a way as to embody and engender fear, the entity so disordered would properly be described as demonic.

The characteristic expressed by the term demonic is complete domination by chthonic drives. In the demon, reason collapses and gives way to instinct. The demon-possessed person has become a purely instinctual being. Emotional drives like hate and fear run rampant. Ordinary people who possess natural instincts are capable of doing things that we might judge as criminal but that is not demonism. The line is crossed when instinct becomes an unchecked force. Testimony to the presence of the demonic takes the form of irrational aggression; sexual crimes, especially those directed against children; hate-filled attacks based on race, sex, nationality, or religion; and senseless acts of vandalism against property—the kind of acts that shock our normal sense of decency. Demonism can also express itself in a fascination with occult phenomenon both on the part of an individual and in a culture as a whole. Demonism is characterized by unreasoning hatred directed against the God whose Word provides our universe with rational structure and moral content. But we must reemphasize the point that not all forms of rejection of the Creator are demonic. Sometimes people just do stupid things.

Life in Bondage to Instinct

The insights we gain from the Day Five narrative in Genesis help us understand what the fall of Adam and Eve meant for humankind and for their universe. The content of Genesis 2 and 3 (the story of Adam and Eve) is not part of this book, but there are some points made in those chapters that are related to the content of Genesis 1, and since our purpose is to understand the message of Genesis I ought to mention them.

Adam and Eve had the opportunity to live their lives on the basis of wisdom (information) that flowed directly and continually from the Creator. When they chose to eat of the fruit that "makes one wise," they rejected that transcendent wisdom to which they had ready access and chose instead to draw their truth from whatever insights might be available within the limited bubble universe they inhabited. This made their thinking subject to the qualities of the material world they had collapsed into existence. That world is necessarily contingent and temporary because it exists only for purposes of measurement and analysis. Once Adam and Eve began to draw their vision of the world from the conditions of material reality, all of reality became for them (not for God) invested with the qualities associated with materiality. Death became for them (and again, not for God) the ultimate fate awaiting everyone and everything. Their minds fell captive to the chthonic forces like fear, hunger, and sexual desire that God had placed in the bubble to protect and disseminate life. These drives were unquestionably good in their role as servants to life, but when they became the driving forces controlling human choices as they did after the fall they proved to be cruel slave masters.

The reason that the fall of Adam and Eve posed such a problem for the whole creation can be stated in terms of quantum physics. As science writer John Gribbon puts it, "The quantum world ... is holistic. ... The world seems to keep all its options, all its possibilities open for as long as possible. The strangest thing about the standard Copenhagen interpretation of the quantum world is that it is the act of observing a system that forces it to select one of its options, *which then becomes real*"[2] (emphasis added).

When Adam and Eve chose to select their options exclusively from the range of experiences defined by observations they had already made, from that point on it was inevitable that every material thing they observed into existence would exhibit characteristics drawn from the bubble-verse itself. Their experience of the world rather than God's schematic became the template they used to draw their new map of existence. Qualities like lust, greed, violence, a drive for dominance, and, ultimately, the omnipresent reality of limitation, contingency, and death would fill every corner of the visible world each time they collapsed part of reality into material expression. This included their experience of their own humanity, which now became in their eyes a purely physical and instinct driven *thing*.

Genesis tells us that the first thing that happened to Adam and Eve as a result of their choice to allow the limits of material reality rather than the wisdom of God to define their lives is that they became aware that they were naked and that they experienced this sense of nakedness as shameful (Gen. 3:7). This sense of shame is revealing. Adam and Eve knew that the way they were generating the physical universe was now out of tune with the will of God. Adam and Eve had been designed to "name all creatures," or as we would say, temporarily collapse the wave structure in accordance with God's blueprint in order to analyze and study God's purposes. But now they had themselves become creatures. The material world that was supposed to be their servant and tool had become their master. The temporary and useful illusion that materiality was supposed to offer now became for them a permanent hallucination and a prison from which they could not escape. It was no longer possible for them to see God as they had done before their fall or obey His will. They had severed their connection to the realm of heaven and they knew it. St Paul expressed the matter perfectly when he wrote, "Through one man sin entered the world and death through sin, and thus death spread to all men" (Rom. 5:12). St. Paul had it exactly right. Death entered the world immediately following man's creation and in precisely the manner Genesis described.

As the biblical revelation unfolds we see how God set to work to limit and then heal the cosmic damage resulting from the fall. God's first move came with the giving of the law. The law was not intended to fix our problem but rather to reveal its existence to us all. By giving us a glimpse of God's original blueprint, the law showed mankind how far it had fallen from God's intentions and how dissonant was the world it was observing into existence from the world God created. The lesson of the law was summarized by St. Paul who, seeing himself in its light, wrote, "I see another law in my members, warring against the law of my mind, and bringing me into captivity to the law of sin which is in my members" (Rom. 7:23).

The law could not fix the problem because it was not possible for human reason to gain control over the persistent illusion that materiality was the only reality. The illusion was just too strong. And yet to concede our fate to the illusion meant the eventual destruction of both mankind and the cosmos we inhabit because of the power of the chthonic forces we had distorted and released. With neither retreat into instinct nor restoration of reason available as options, a third path to the salvation of the world had to be found.

Mankind needed to be freed from our self-made snare. We needed to be cleaned up and given a fresh start. We could not do this on our own so God sent His own Son to pay the penalty of Adam's sin. With His death the opportunity was made available to begin our lives all over again with a clean slate. All God asked any of us to do was to believe that the sacrifice had been made and accept the forgiveness won at the cross.

William Blake wrote, "If the doors of perception were cleansed, everything would appear to man as it is, Infinite."[3] God's forgiveness cleanses the doors of our perception. The love God bestowed on us restores communion with God for all who believe and starts our lives over the right way. Jesus described God's plan in an interesting way. He said, "Unless one is born again he cannot *see* the Kingdom of God" (John 3:6 emphasis added). Note the importance of *seeing* the Kingdom. By giving His life, Jesus made it possible for us to see again. The thing that we see, by faith, is the original blueprint of God, the wisdom and eternal pattern that was supposed to guide our lives from the beginning.

Set beside instinct and knowledge, faith is a third way of knowing. Faith makes instinct subject to knowledge because faith has rational content, the Word of God presented in the Bible. Minds informed by the information in that Word can collapse the wave structure in ways a little more consistent with the will of God and so begin the transformation of the cosmos. But faith is not sight and that too is important. People of faith are not perfect. We do not always do the right thing because we do not always see our way forward clearly. Faith in the sacrificial death of Jesus Christ begins to restore our rightful mind and move us back toward our proper place in the universe but our healing will not be complete until faith becomes sight and we stand in the presence of God. Until then, all we can say is that while Christians share the same world their neighbors live in, they are no longer enslaved to the realm of materiality and the drives that master it. They live today in the proleptic light of a new heaven and new earth in which the universe and the humanity that God planned in the days of creation, a reality now seen only dimly and sporadically, will fill our vision forever.

Chapter 9

Ecce Homo: Genesis Day Six

As Day Six dawns in the text of Genesis we enter a new world. The language of the text becomes richer and more interesting. The text itself is more complex and it is longer. But the subject is the same.

Genesis is describing the creation of the cosmos, an event that took six twenty-four-hour earth days. The event described in Genesis has nothing to do with the physical construction of our universe (with, as we will see, one important exception). Genesis describes the encoding of our universe; it outlines the information that will determine the shape of the material universe when it comes into existence.

There is no other way Genesis could have presented its topic. Thanks to quantum physics we now understand that the building blocks of our universe are mathematical probabilities that exist in the form of nonmaterial energy states until a human observer transforms those energy states into physical particles by an act of observation and measurement. These physical particles are the foundation upon which all material entities are built.

America's greatest physicist, John Wheeler, once described the quantum approach to physics as the "It from Bit Theory." Michio Kaku outlines Wheeler's presentation of quantum physics in words that should immediately bring the worldview of Genesis to mind:

> [It] starts with the assumption that information is at the root of all existence. When we look at the moon, a galaxy, or an atom, their essence, he claims, is in the information stored within them. ... The universe ... sprang into being because it was observed. This means that "it" (matter in the universe) sprang into existence when information ("bit") of the universe was observed. He calls this the "participatory universe"—the idea [is]... that our very presence makes the universe possible.[1]

This way of thinking about things may sound strange when presented as a scientific theory but in real life we live in the world of "It from Bit" every day. Every entity is either a physical object that we are seeing, hearing, or touching or else it is an idea that has the potential to become a physical object. If you do not in some way observe me, I am for you information that exists as an idea in your mind. If you encounter me on the street then and only for the length of that encounter am I a physical reality for you. When our encounter ends, I revert to the status of information (a memory, picture, description, etc.).

The only new idea at issue in the "It from Bit" approach involves the question of which form of existence, ideational or physical, Bit or It, is primary? Old-style materialism assumed that the definitive reality of anything was its physical expression. Adherents to this nineteenth-century philosophy viewed all forms of information as deriving from a physical encounter. For them, information was only the product of the observation and measurement of the physical qualities of an object. Quantum physics does not eliminate the distinction between idea and physical expression, it simply reverses their order. It tells us that the informational existence of anything (in the form of the mathematics of its preset energy levels or quanta) is the normal and natural condition that precedes its physical expression and that any physical forms that might appear are second-level phenomena contingent upon the presence of a human observer actively taking measurements. As a result of this order of appearances, material reality is always secondary and artificial in the sense of being an artifact resulting from observation. Niels Bohr, the founder of quantum physics, put the matter very clearly when he said, "An independent reality in the ordinary sense can neither be described to the phenomena nor to the agencies of observation."[2]

This may all sound cutting edge but eighty years have now passed since the founders of quantum physics met together and published the Copenhagen Interpretation that defined this new science. There is no doubt that over the coming century every one of us is going to have to come to terms with the way in which quantum physics reorganizes reality. The imminent advent of quantum computing, where computational power will not be limited by any spatial considerations because the computer will use energy fields rather than physical particles to do its work, will force people to begin to incorporate this view of reality into their daily thoughts.

I believe that this way of looking at reality is something we can learn to like. Quantum physics invests human beings with greater dignity than materialism can offer. Materialism by definition views human beings as just another part of the physical world. The materialist assessment of human life was summarized by evolutionist Jean B. De Lamarck, who wrote, "Life is merely a physical phenomenon. All manifestations of life are based on mechanical, physical, and chemical causes, being properties of organic matter."[3] Quantum physics, and, as we are about to see, Genesis itself views humans as the progenitor of physical reality. The difference between the two approaches was beautifully stated in an important observation by America's most influential philosopher, Alfred North Whitehead:

> Nature gets credit which in truth should be reserved for ourselves, the rose for its scent, the nightingale for his song, and the sun for its radiance. The poets are entirely mistaken. They should address their lyrics to themselves and should turn them into odes of self-congratulation on the excellence of the human mind. Nature is a dull affair, soundless, scentless, colourless, merely the hurrying of material, endlessly, meaninglessly.[4]

Writing in the days when philosophical materialism still dominated intellectual life, American philosopher Ralph Waldo Emerson offered an assessment of the importance of human beings that anticipated Whitehead when he observed that "We animate what we see and we only see what we animate."[5] Centuries before either Emerson or Whitehead, Galileo intuited precisely the same insight when he concluded that "tastes, odors, colors, and so on [are] no more than mere names so far as the object in which we place them is concerned, *and they reside only in consciousness*"[6] (emphasis added).

The Bible has always warned us that we tend to sell ourselves short because we see ourselves differently from the way we really are. This was his point when St. Paul wrote, "For now we see in a mirror dimly, but then face to face. Now I know in part, but then I shall know just as I also am known" (1 Cor. 13:12). St. John expressed this idea of a difference between the way we presently experience ourselves and the truth of who we are but expanded it to include our experience of Jesus when he wrote, "Beloved, now we are children of God; and it has not yet been revealed what we shall be, but we know that when He is revealed, we shall be like Him, for we shall see Him as He is" (1 John 3:2). Jesus' brother James underscored that point when he described someone who hears the Word of God and does not do it in terms suggestive of the quantum understanding of the universe: "He is like a man observing his natural face in a mirror; for he observes himself, goes away, and immediately forgets what kind of man he was" (James 1:23–24).

Spiritual Information

In this important verse James is telling us that there is a significant difference between the natural man as he appears to himself in the wave-collapsing moment of self-observation and the person he really is, a reality that is forgotten by most of us most of the time. According to James, while the sensual observations made by mankind may sometimes inform our vision of ourselves, they can also confuse us. The key to the verse is the words *natural man*. Throughout the Bible, this phrase never describes humanity as seen in the eyes of God. These words are connected to a false and limited perspective. They are usually contrasted to the notion of the "spiritual man." Spirit, as we have seen, refers directly to the realm of *logos*, *aidos*, and *hokma*, what we are calling the world as information. James is telling us that observation causes us to forget "what kind of man" we are by diverting our attention from the realm of spirit/information, which contains the eternal content of our lives. We tend to focus instead on the conditional and temporary level of materiality.

There is nothing wrong with material reality. The problem with defining ourselves in material terms rests in the fact that physical reality is always a secondary, artificial, and contingent condition. We can only define our lives accurately in the light of their primary and enduring form. The Bible presents a very positive vision of human possibilities (and thereby a very tragic assessment of our failures) because it frames our lives in terms of their informational rather than physical content. In the gospel of Matthew, Jesus speaks of a kingdom that has been "prepared for you from the foundation of the world" (Matt. 25:34), implying that we have existed as knowledge in the mind of God from the beginning of creation. Reflecting the same point of view, God said to the prophet Jeremiah, "Before I formed you in the womb I knew you; before you were born I sanctified you, I ordained you a prophet to the nations" (Jer. 1:5). God is telling the prophet that he existed in the form of a plan in God's mind long before the name *Jeremiah* was ever connected with any material reality. Psalm 139 offers a vision of humanity that finds an uncanny echo in quantum physics when the psalmist writes, "Your eyes saw my substance being yet unformed. And in your book they all were written, the days fashioned for me, when as yet there were none of them" (Ps. 139:16). The phrase *in your book* implies preexistence in the form of information. These verses all locate the origin of our lives not in physical events but in a blueprint designed before anything material came into existence, the blueprint described in Genesis.

The biblical point of view and its quantum echo allow us to think credibly about our lives in an eternal dimension that stretches backward in time but it also leaps forward into the future. This is what the Bible is promising when it speaks of resurrection. The gospel tells us that we will be raised from the dead just as Jesus was raised from the dead. The notion of resurrection seems problematic and maybe even a touch macabre if materialism is right and all we are is a bag of protoplasm. In that case, resurrection would seem to involve the reanimation of a decaying corpse and who needs that? But if mathematically stated information is the true and natural condition of our existence, as both the Bible and quantum physics say, then the picture changes.

Think of the book you are holding in your hand. Under the conditions of the bubble-verse, that book has been observed (by you) and its information collapsed into a material form. You hold that form now. But I can assure you that the book does not actually exist in that form. The book exists in my mind. The authentic form of its existence is as information. This means that the book is capable of resurrection. You can toss the book away or even burn it but that will in no way eliminate the information. The information still exists in my mind, and besides that, it persists in the hard drive of my computer where it takes the digital form of a set of zeroes and ones. Even if I disappear, that digital information can be reconstituted any time. We can now extend that analogy. If we exist *primarily* as information, the idea of resurrection becomes not only credible, it becomes a virtual certainty for the reasons we just described. The bottom line is that we are sometimes physical bodies *but we are always information* and that is what makes resurrection a certainty just as the fact that this book exists primarily as information makes its restoration after any form of destruction both easy and inevitable.

The fact that we are not limited by the material qualities of the body does not, however, mean that eternal life is automatically ours. There is a potential limit to our existence even as information. Our lives can be taken away permanently if the information that we are is destroyed. This is what Jesus was talking about when He urged us to keep our focus on protecting the quality of the information that provides the enduring reality of our life: "And do not fear those who kill the body but cannot kill the soul. But rather fear Him who is able to destroy both soul and body in hell" (Matt. 10:28).

We can understand that verse very well at this point in history. Our existence as some form of body is assured as long as our life as information is intact. We don't need to worry about the one (Satan) who presently manages the chthonic forces in the material realm or about the physical form of ourselves that for the moment may be subject to his influence any more than an office worker needs to print and save a physical copy of every personal e-mail she has received. But we do need to concern ourselves with the one (God) who has His finger on the delete button. Physical copies of e-mail can always be restored—unless the information has been trashed and the trash is purged. It is the same with any particular physical manifestation we may express. The loss of a physical copy does not entail any loss of information. The Bible does, however, warn us that those who have unreservedly identified themselves with their physical life in this bubble-verse will share its fate. When the bubble-verse is destroyed, they will go with it.

As an aside, it is interesting to notice how easy it is to use imagery drawn from the use of computers and cutting-edge physics to illustrate the message of the Bible. This is a phenomenon well known to biblical scholars. As our scientific knowledge and technology advance, the way we experience the world begins to resemble the language and assumptions of scripture more and more closely and the Bible becomes progressively easier to comprehend. The book of Revelation, for example, presumes that when Jesus returns to Bosra (today's Petra) in Jordan all the people of the world will see Him. For centuries this promise made no sense. Today, in a world where satellite news instantly brings events to every nation, nothing in that promise suggests anything miraculous anymore.

This increasing correspondence in worldviews should not surprise us. The world is as the Creator made it. Ideas rooted in the truth of the Creator's work, no matter how ancient, will necessarily reappear throughout history. As Aristotle told us long ago, "It is not once or twice but times without number that the same ideas make their appearance in the world."[7] Julius Robert Oppenheimer,

Director of the Manhattan Project that built the atomic bomb, offered the same observation: "The general notions about human understanding which are illustrated by discoveries in atomic physics are not in the nature of things wholly unfamiliar, wholly unheard of, or new. Even in our own culture they have a history. … What we shall find is an exemplification, an encouragement, and a refinement of old wisdom."[8]

Today the science of physics is returning to a path hewn by the ancients. After decades spent wandering in the darkness of materialism, we have begun to understand our universe as a light-driven system for communicating mathematically stated information. The impetus behind this renewal is the rediscovery of the insight expressed by Isaac Newton who observed that "the entire universe is a cryptogram set by the Almighty."[9] The thing we have not done so far is to go all the way back and take a look at the actual code that is the original content of the universe we inhabit, the code recorded in the book of Genesis. That mistake is being corrected now.

We have discovered that human life stands at the crossroads of information and materiality. Reality remains mathematical probability until a human being transforms probability into particles and so brings the material world into existence. The design of an entity who is capable of doing that is therefore one of the most significant acts not just in the story of life but in the history of the universe. The material expression of everything that exists will depend on the character and qualities possessed by the entity that collapses mathematics into physical form. How s/he observes the world will make the information coded as potential during the first six days of history tangible. Let us turn, then, to the study of man, as presented in Genesis. *Ecce homo*, the title of this chapter, expresses its theme: Behold the Man.

Our Number is Six

In Genesis, reality consists of a set of the numbers one through seven and, as we will see, various combinations of those numbers together with corresponding geometric figures. The key to understanding humanity is to understand the number six, for this is our number. The Bible could not be clearer on this point. Psalm 90 is a poem penned by Moses himself. It expresses the thinking of the man who, directed by the Holy Spirit, gave us Genesis. Speaking through that poem, Moses gave us this advice: "Teach us to number our days that we may gain a heart of wisdom" (Ps. 90:12). In some of the most ancient words in the Bible, Moses told us that if we want to gain wisdom, we would do well to pay attention not to our days but to their number. I believe he offered this advice as a key to deciphering the text of Genesis.

Centuries later while looking forward toward the end of history from his cave on the Isle of Patmos, St. John wrote this in the book of Revelation: "Here is wisdom. Let him who has understanding calculate the number of the beast, for it is the number of a man: His number is 666" (Rev. 13:18). The number of the beast is six amplified, six being the number of humanity. John's message is clear; know the number, know the man. So the question about ordinary man becomes a question about his number. What is six?

There are really only two things we need to know about six. First, six is the sum of other numbers. Six can be the sum of two plus four. As we have discovered, two is the principle of conscious awareness. Four is the number of our completed four-dimensional universe. The combination of two and four means that through the number six the universe becomes reflexively self aware. This set of numerical relationships defines a human being as the point where the universe becomes conscious of itself. This is another way of saying that it is through the wave-collapsing activity of human consciousness that material reality arises, giving the nonmaterial realm of information visible expression.

Six can also be formed by combining one plus five. As we have seen, one is not a number but a concept, the monad or singularity. It designates the core or constant around which the universe is gathered. In the Bible (and in Einstein, of course) that constant is light. We have seen that five is a dark number expressing the instinctual or chthonic character of existence. The combination of one and five is man in a fallen state. This is the decrepit condition of the being we described earlier who was made to transform information into materiality but who has become trapped in the physical world he himself has generated.

Six can, however, also be formed by combining three and three. Three is the life generating number. The combination of three plus three speaks of a humanity that gives life to the bubble-verse we inhabit. This is the form of humanity Alfred North Whitehead celebrated that increases the vitality and abundance of the world. In this equation, man enhances the world as he gathers its informational potential to material expression. He is not at odds with its mathematical potential—quite the contrary. By collapsing the wave structure of the bubble-verse in the right way, he enables the universe to realize its fullest possibilities. This is man as he was originally created to be.

Very similar to that vision of humanity is man formed by the combination of one plus two plus three. As we have seen, three is the principle of life, two represents the concept of conscious awareness, and one is the organizing principle of light that provides the medium for the communication of information. In this arrangement, all the elements of man's nature work together to give expression to the creative intentions of God. This is the man Jesus spoke of when he said, "You are the light of the world" (Matt. 5:14), a phrase that describes the community of the redeemed operating at its best.

To summarize, the first thing to notice about the number six (man in his informational essence) is that six is a combinational number full of creative potential that can generate many different kinds of universes depending on which combination formed the number.

The other thing to notice about the number six is that it is not seven. For Pythagoreans, seven was a very special number. It was, for example, the number of the completed musical scale. But it was far more than that. Seven was not seen as a combination of numbers but was treated as itself a mysterious power. It represented a point of contact between something that was part of our reality (it was, after all, a number) and something beyond our experience. Because of this, seven was the number that expressed the presence of the Creator in His relationship to His creation. I do not intend to explore the number seven in this chapter; for now all we need to see is that six is definitely not seven. Seven is sure and consistent in its meaning. Six is pretty shaky. It might go this way or it might go that. Seven is perfect and six is close, but this is cosmology not horseshoes, and close is not, as we will see in both Genesis and human history, good enough.

We are now in a position to easily understand the Genesis Day Six narrative because that narrative is a verbal explication of the problems and principles we have discovered in our quick analysis of the number six itself.

God Did the Math

The text of Genesis Day Six begins this way:

> Then God said, "Let the earth bring forth the living creature according to its kind: cattle and creeping thing and beast of the earth, each according to its kind," and it was so. And God made the beast of the earth according to its kind, cattle according to its kind, and everything that creeps on the earth according to its kind. And God saw that it was good.
>
> —Gen. 1:24–25

The first thing to notice is that these verses confirm our proposal that the focus of Day Five rested on the encoding of chthonic forces rather than the creation of specific living creatures. The Day Six text describes the creation of information that encodes the full range of life forms. The text mentions beasts and cattle but also "creeping things," a term that appears on Day Five but which must in that context have been referring not to a specific creature but to chthonic forces in general, just as we suggested. The Hebrew term refers not only to "creeping things" but also to creatures that glide through the sea and all animals that have the ability to move around. The word rendered as *cattle, b@hemah,* literally means "mute" and is an expression collective of all land animals domestic and wild. The terms used in the Genesis Day Six text tell us that this is the day on which the mathematics of every kind of creature on earth was encoded.

The second thing to notice is that once again it is the element Earth that brings forth this code. In this portion of Genesis, God is passive. He permits Earth to "bring forth" creatures but He Himself does not actually do anything. The Hebrew term rendered as *bring forth* is *yatsa.* It expresses the idea of "bringing something out of oneself." In this case, the self that brings forth all these creatures is the element Earth. In our discussion of the Genesis Day Three text we explained why it is in both Genesis and subsequent Pythagorean thought that the number three and its associated element Earth contain all the information necessary for life. The arguments we set forth in that chapter are confirmed by the fact that the generation of the code of life is here completed precisely on Day Six, a day whose number is the number three amplified.

The Day Six text of Genesis centers around certain key words and we will use these to guide us through the entire text. The first of these terms is the Hebrew word *nephesh.* In translation this word is often rendered as "living creature" but this rendering is somewhat inaccurate. *Nephesh* actually communicates the idea behind English words like soul, self, inner being, and activity of mind. It also refers to the ability to feel, act, and will something.

This word is joined in the text with another word that is often translated as "beasts of the earth." Here the Hebrew term is *chay.* This word refers to anything that is alive, including vegetation. The appearance of these two words on Day Six, terms that embrace the full range of living beings, permits us to say exactly what was coded into the bubble-verse on Day Six.

The product of Day Six was a digital code, specifically the DNA code. The text describes this four-digit code in terms that reflect its importance and its capabilities. The DNA code contains sufficient information to provide the data that can express itself in the form of every kind of creature from the simplest vegetation to the most complex sentient being. Current research indicates that it may also contain the information that organizes many behavioral traits and mental operations observed in living creatures.

Our conclusion that Genesis refers here to DNA is confirmed by the text's repeated insistence that everything on Day Six is being brought forth "according to its kind." The Hebrew term rendered as *kind* is *miyn*, a word expressing the English concept of species with particular reference to what we today identify as the gene pool that distinguishes one species from another in reproduction. This distinctive genetic content that plays such a crucial role in the preservation of species from generation to generation can, again, be nothing other than the digital code we call DNA.

It is important that we do not slip into the lazy habits of materialism at this point. The DNA generated on Day Six in Genesis is entirely a mathematical code. When the wave structure of pure probabilities is collapsed, this DNA code will certainly take the form of physical chemicals arranged in a double helix, but that is not what is created on Day Six. The DNA code written on Day Six is different from any shadow expression of that code that might subsequently appear in the chemistry of the physical world. This reflects the difference between paper and message that we described earlier. Materiality is the paper on which information sometimes appears. Materialist science scrutinizes the qualities of the paper because, given its myopic focus on physical reality, it must assume that somehow the paper is the source of the message printed there. But as German scientist J. Peil reminds us, "Information is neither a physical nor a chemical principle like energy and matter."[10] His point is that matter cannot write codes; the only kind of information we know anything about has in every instance been generated by minds. Since our universe is a system of communication rather than a collection of things, to understand it we must seek the source of the information that is being communicated. This is what Genesis does. It is telling us where the information coded into DNA comes from. God coded into the element Earth (expressed geometrically as a cube) a sequence of digits that could, when observed and measured by human beings, take expression by means of DNA as the population of living creatures that fill the universe.

With the coding of the sequences of DNA on Day Six the text of Genesis has reached a point where everything needful for the universe is present except for one piece. What is needed now is a creature who is capable of turning all the mathematical potential we have described into physical reality, a creature who can by its observations and measurements, literally its "naming of everything" as Genesis 2:19 puts it, collapse the sea of probability waves that Genesis has set churning and give to God's universe of endless possibilities physical definition and material structure. In the text of Genesis, the universe sits silent for a moment, awaiting the creation of man.

We Work the Problems

Genesis describes the creation of man in these words:

> Then God said, "Let us make man in Our image, according to Our likeness, let them have dominion over the fish of the sea, over the birds of the air, and over the cattle, over all the earth and over every creeping thing that creeps on the earth." So God created man in His own image, in the image of God He created him, male and female He created them. Then God blessed them, and God said to them, "Be fruitful and multiply; fill the earth and subdue it; have dominion over the fish of the sea, over the birds of the air, and over every living thing that moves on the earth."
>
> —Gen. 1:26–28

This is obviously a very complex text with material enough for several books. What I want to do here is to lay out the bones of the text, the basic structure, by continuing to focus on just the most important words.

The first word applied to man is *image*, *tselem* in Hebrew. Theologians and biblical scholars have debated the meaning of this term for centuries. Much of that debate is irrelevant because it has been couched in materialist terms. Various scholars have tried in many ways to define the image of God by identifying some point of physical similarity between man and God. None of these efforts commands universal support for the obvious reason that it is impossible to make a case for any material similarity between man and God.

What these scholars have overlooked is that the universe we inhabit is not a collection of objects but a system of communication. In that context the word *image* assumes a specialized meaning. In a system characterized by the communication of information, an image is a means for transmitting data. A map, for example, is an image that conveys information from one mind to another. Photographs and computer generated images do the same thing. In the context of a system designed to communicate information, the word *image* refers to the syntax by which such communication takes place.

The "image of God" that Genesis speaks of must be understood in this way. The phrase cannot refer to an objective similarity between man and God because in the Bible neither man nor God is considered an object. God is known through His *logos, aidos,* and *hokma* and man is the lynchpin that connects the mind of God to the forms of physical expression that God's thoughts will assume. The phrase *image of God* is telling us that man is a syntactical device God uses to translate the mathematics of creation into material form.

The Hebrew term *tselem* actually renders the English terms *shadow* or *shade.* It refers to something empty. The word can even render the English term *mere.* All these definitions taken together confirm that mankind is the ghost in the machine that makes the thoughts of God visible. Man is a medium of communication between the eternal and temporal and in that sense is the image of God. He is not a miniature version of the Lord.

It is important that we understand the meaning of the phrase *image of God.* In the Bible, mankind is continually threatened by the material world we have generated. The danger we face is that we might forget that we are its progenitors and not its product. In the Bible, the materialist fallacy is connected with the word *idolatry*. This is also a concept that is often misunderstood.

The Hebrew term rendering the English word *idol* is the same term we are now analyzing in Genesis, *tselem,* meaning a shadow or shade. Just as the concept of the "image of God" has nothing to do with physical resemblance but refers instead to man's role in the communication of God's will, in the same way the word *tselem* when applied to the idea of idolatry has to do with a breakdown in the realm of communication.

The main problem with idolatry, as the Bible sees it, lies in the fact that idolatry elevates material reality to an importance it does not deserve. Materiality exists so that human beings can perceive and analyze the thoughts of God. It does that by making those thoughts palpable. Physical reality serves its purposes when, through careful study including scientific research, human beings are able, as Einstein put it, to glimpse a bit of "the mind of God." Idolatry is the fallacy of treating a means of communication (materiality) as if it were the source of the message and an end in itself. This error is similar to a child's belief that the figures s/he sees on television actually inhabit the set. The delusional thinking at the root of idolatry was embraced by German evolutionist Manfred Eigen when he offered this mindlessly self-contradictory assertion: "The *logic* of life originates in *physics and chemistry*"[11] (emphasis added). Eigen seems unaware of the fact that while chemicals and physical forces can bear witness to a source of rational order, no chemical or elemental particle possesses in itself the quality of logic.

In a passage that explains the real issues involved in idolatry, the prophet Isaiah begins by condemning Eigen's kind of thinking. The prophet tells us that "those who make an image, all of them are useless" (Isa. 44:9). The idea is that those who see human beings as products of the physical forces they themselves set into motion have nothing useful to say to the rest of us because they do not understand how the world really works.

To help us understand why this is so, Isaiah brings us into an ancient wood shop and describes how an idol is made: "The craftsman stretches out his rule, he marks one out with chalk; he fashions it with a plane, he marks it out with the compass, and makes it like the figure of a man. … He cuts down cedars for himself and takes the cypress and the oak; he secures it for himself among the trees of the forest. He plants a pine, and the rain nourishes it" (Isa. 44:13–14).

The prophet's words then become satiric and his tone a bit sarcastic as he continues his description of the process of making an idol: "He makes a god and worships it; he makes it a carved image, and falls down to it. He burns half of it in the fire; with this half he eats meat; he roasts a roast, and is satisfied. He even warms himself and says, 'Ah! I am warm, I have seen the fire.' And the rest of it he makes into a god, his carved image. He falls down before it and worships it, prays to it and says, 'Deliver me, for you are my god!'" (Isa. 44:15–17).

Isaiah is showing us that the problem with idolatry does not lie in the object that is produced but rather in the thinking that produces it. The craftsman's actions in making an idol are offered by the prophet as an illustration of something we all do every single day taken, in his example, to such an extreme that we can, for perhaps the first time, see its folly. By our observations and measurements we create this artificial material world for our convenience ("to bake our bread"). Isaiah reminds us that it is the craftsman himself who "plants a pine," "cuts down cedars," "marks it out," and "fashions it with a plane." We give form to every physical expression reality assumes by the observations and measurements we make, but then we imagine that this world we just made has somehow made us. Isaiah stands dumbfounded before the folly of such a deluded craftsman.

There can be no more precisely targeted critique of the materialist fallacy and the error of evolution. Those who advocated these nineteenth-century ideas simply reversed the way we now know the universe works. They claimed the product (physical reality) was the maker and said that the maker (the human being) was the product (of physical reality). In Isaiah's view, those who so reverse reality and who teach others to do the same not only have fallen into error, they actually stand condemned before both God and man: "They do not know nor understand; for He has shut their eyes, so that they cannot see, and their hearts, so that they cannot understand… He feeds on ashes; a deceived heart has turned him aside; and he cannot deliver his soul, nor say, 'Is there not a lie in my right hand?'" (Isa. 44:18, 20).

The only way to protect ourselves from the idol makers among us is to remember that the world we see, to whatever degree we see it, hear it, speak about it, or write about it, is to that degree "a lie in our right hand." It can be magnificent, pleasurable, and useful but it is not in any ultimate or final sense real. Physicist John Wheeler, no died-in-the-wool materialist, expressed his relationship to the material world beautifully:

> I confess that sometimes I do take 100 percent seriously the idea that the world is a figment of the imagination and, other times, that the world does exist out there independent of us. However, I subscribe wholeheartedly to those words of Leibniz, "This world may be a phantasm and existence may be merely a dream, but this dream or phantasm to me is real enough if using reason well we are never deceived by it."[12]

The physical world is the field in which we were born to spend our lives. We are supposed to respect and care for it. But the material world is potentially deceptive because, while it appears to present the "bottom line" and "hard facts" of life, nevertheless it is only the contingent expression of a reality whose eternal truth exists only in the form of mathematics and energy states. Jean Jacques Rousseau unwittingly summarized the message of Isaiah very well when he said, "Nature never deceives us, it is always we who deceive ourselves."[13]

The first word Genesis uses to describe humanity, *tselem*, in English *image*, is a word full of deep implications that require our careful consideration. In the context of a universe that consists in the communication of information the word *image* reminds us that we are appointed by God to be the means by which His blueprint for creation is translated into material forms. This word protects us from over-identification with the world we spin around ourselves. At the same time, it challenges us to engage in our portion of the Creator's task with diligence. We have been designed to turn ideas into landscapes, cities, and technological wonders but we must do this with hearts prepared to listen for the will of God and bring that will to fruitful expression.

The Potential Within Patterns

The second word to which we need to give attention is rendered *likeness* in the New King James Bible. The text tells us that man is formed in the "likeness of God." The Hebrew word *d@muwth* expresses the idea of similitude. The key question is to what kind of similarity does this word refer?

We have seen that the Bible views our world as a system of communication rather than a collection of objects. The world began as Word and remains the expression of Word. It is essential that we keep this in mind as we examine each of the words used to describe humanity in Genesis. In this case, the term *d@muwth* tells us that a copy of God's blueprint for creation has been planted in our minds.

It is easy to demonstrate that the Bible uses this particular term to refer to a schematic. In 2 Kings, for example, we see that the term has a clear technical meaning: "Now King Ahaz went to Damascus to meet Tiglath–Pileser king of Assyria, and saw an altar that was at Damascus; and King Ahaz sent to Urijah the priest the design of the altar and its pattern *(d@muwth)*, according to all its workmanship" (2 Kings 16:10).

In this passage, the word *d@muwth* obviously refers to a blueprint. King Ahaz has observed and measured an altar and is relaying its dimensions to artisans in Jerusalem by means of a *d@ muwth* or schematic. It is important to notice that this blueprint is designed and disseminated in the context of communication. In a culture such as ours that has labored under the presumptions of materialism for a century and a half, it is easy for us to assume that the word *likeness* refers to some kind of physical similarity between two objects. But *d@muwth* clearly describes the content of a communication.

Genesis' use of the term *d@muwth* tells us that the information that is the essence of who we are contains part of the information that is the essence of who God is. In other words, a part of the blueprint of creation has been fixed in our minds in such a way as to predefine parameters for how we will collapse the wave structure of unlimited possibilities and generate the material world. This is a point where Genesis makes another contribution to quantum physics.

We have seen that the law of the conservation of matter and energy requires that every potentiality existing in the quantum realm must come to expression. This requirement has forced quantum physicists to insist not only that everything is possible (the sun can set in the east, for example) but that every possibility will necessarily find physical expression at some time or another. This is so because any unexpressed possibilities (which exist as preset energy fields) will have been irretrievably lost, an impossible event in light of the requirement that energy can neither be created nor destroyed. Up until now, this aspect of the theory has provided an easy point of attack for opponents of the quantum approach to reality. We have also seen that the Bible offers a way to resolve this problem by identifying the life of Jesus Christ as the point where all possibilities come to expression, including those that never receive physical manifestation.

The word *d@muwth* denotes a second means by which the commonsense flow of history is preserved.

Genesis informs us that a copy of God's blueprint has been planted in our minds. This means that we are not able to conjure just any world into material existence. The universe we observe into existence for purposes of analysis and measurement must remain within parameters established by the blueprint penned by God and coded into our minds. This guarantees that our world will tend to be reasonable and predictable. I find it ironic that many biologists still insist that the methodological atheism characteristic of materialist science and the randomness demanded by the theory of evolution provide the foundation for all scientific work when the fact is if everything really was the product of unpredictable accidents and spontaneous self-generation scientific research would be impossible. It is Genesis with its insistence that there is a blueprint behind creation, a blueprint copied into the mind of humanity, that provides a stable foundation for scientific research. It is no accident that science has gained a permanent foothold and flourished only in those cultures for which Genesis has provided the traditional starting point for cosmology and in societies that have come under the influence of Western thinking.

The presence of a blueprint etched into the mind of man also tells us why certain events in history are presented in the Bible as preordained. When Jesus was preparing His disciples for His imminent death, He told them that "all things that are written by the prophets concerning the Son of Man will be accomplished" (Luke 18:31). Throughout the Bible, certain human decisions and actions are described as "necessary" or as occurring "at the appointed time." This language tells us that while many human choices seem to be a matter of free will, some events are not. History appears to operate like a log flume ride at a theme park. For long sections of the ride, the log just bounces from side to side within a fairly wide channel. But at certain points, such as the final lift, the log is suddenly fixed to a rail like a roller coaster and the range of motion is strictly controlled. Events connected with Jesus, especially His first visitation and final return, seem to be precisely determined as demonstrated by the detailed prophecies that surround those events. The blueprint God planted in the human mind that normally offers us a range of choices is also capable of precisely dictating our actions when the will of God demands that.

The next word in Genesis that offers insights into the nature of man is the Hebrew term *radah*. The word is translated in the New King James as *dominion*. The Hebrew word actually has sharper and more aggressive connotations than would be suggested by that translation. Its meaning is more accurately rendered by such English words as *subjugate*, *tread down*, and *scrape out*. In its normal usage the term refers to an exercise of political authority by which a person or group has subjected a certain population to tyranny (Isa. 14:2).

The word *radah* is paired in the text of Genesis 1 with the term *kabash*. The pairing occurs between verse 26, where God says, "Let them have dominion *(radah)*" and verse 28, where God says, referring to the creature he has just encoded, "Fill the earth and subdue it *(kabash)*." *Kabash* like *radah* describes an act of taking possession of something but emphasizes the fact that this is being done in a violent way and tells us that the object in question is now subject to bondage. It can be rendered by English terms like *tread* and *trample*, and in the book of Esther it refers to a rape (Est. 7:8).

Once more we must remind ourselves that the context in which these words are to be applied is a universe that is a system of communication rather than a collection of objects. In Genesis these words are not intended to define the character of a human being. In the context of communication they refer, rather, to the task of translating God's plans into physical form. In that context, the significance of the words does not rest in their violent and aggressive connotations but rather in their assertion of absolute authority. This passage never intended to describe *how* mankind would rule the physical world; instead it tells us that the power God has given human beings to determine the form of that world is unchallenged and unequaled and should be exercised with bold creativity. Genesis uses the language of despotism because that language establishes the authority of humanity using images that could be understood by all people in all ages. Not every culture has experienced democracy or a republic but everyone knows what despotism is. The point of the passage is that no other creature coded into the nascent creation possesses anything like the human capability to collapse the wave function of pure potentiality and bring part of that potential to physical expression. These words and their field of meaning offer nothing less than a summary of the Copenhagen Interpretation of quantum physics, which states that there is no physical reality independent of a human observer.

The words *kabash* and *radah* are used in Genesis to remind us that our mastery over the material realm must never be compromised or surrendered. Should that happen, we would lose our unique human identity. This is exactly what has befallen evolutionists. Evolutionists have lost sight of their role in creation to the point where various swamp creatures have become for them totemic ancestors and even creator gods. Believing that they are nothing more than the blind products of physical and chemical forces, they have drifted into a primitive form of animism. The oddity of our time is that once again while these folks would have us believe that by advocating such notions they are fighting to advance reason, enlightenment, and "science" and so save us all from another "dark age," the real fight to save humanity from descent into the realm of the beast is being waged by their critics.

No Two Alike

The actual encoding of humans is described in Genesis 1 by the Hebrew word *bara*. The term is often translated *created* but actually refers to an activity more akin to manufacturing. It is best rendered by the English words *to shape* and *to fashion*. The term literally describes factory operations like cutting, paring, polishing, and, interestingly enough, planning, as in making a blueprint.

All these connotations of the term *bara* constitute a claim that for Genesis a human being is a unique creation to which God gave personal attention. Normally in Genesis the element Earth brings forth living creatures, but this not so in the case of man. The blueprint for man is drawn up personally by God and then, as we will see, it is God who personally goes on to handcraft the physical form of a human being. This is so because man is too important to the future of the universe for his creation to be left to the operations of what we would call natural laws. The code that we are and the form we take are the product of a process that we can safely call intelligent design.

Throughout this book I have tried to communicate the fact that many hundreds of critics within the scientific community are urging biologists to move beyond their myopic attachment to nineteenth-century Darwinism. Among these critics is Dr. Francis Collins, director of the Human Genome Project, one of the monumental scientific achievements of our age. His approach to the issues involved with origins is similar to that adopted in this book. He believes that much of natural history can be accounted for by appeal to natural laws, a process we are calling *derived naturalism*. But he is not convinced this is true in regard to the very beginning of the process of creation, where the conditions that permitted life seem to have been rather carefully designed, and he sees the direct hand of a creator revealing itself again in the design of human DNA. He writes, "Humans are ... unique in ways that defy evolutionary explanation and point to our spiritual nature."[14]

Dr. Collins's comment effectively summarizes the claims of the Genesis text. In Genesis, God is unusually hands-on in the design of human beings. Genesis 2 describes how God gave human beings their unique physical shape: "And the Lord God *formed* man of the dust of the ground, and breathed into his nostrils the breath of life and man became a living being" (Gen. 2:7, emphasis added).

It is important to pay attention to the phrase in which we are told that God "formed" man. That phrase tells us that the physical form, including the brain and nervous system characteristic of humans, was uniquely built by God himself and was not a product of nature. The reference to God infusing into man the "breath of life" tells us that God also specifically designed the human DNA code and then inserted it into the physical being He had formed. We know this because the phrase *breath of life* is in Hebrew a reference to the same Holy Spirit who brooded over the creation at the beginning of Genesis, a Spirit identical with the *logos*, *aidos*, and *hokma* residing from eternity in the mind of God. The phrase *breath of life* therefore must refer to the informational content of a living creature or, in other words, to its DNA. Genesis says nothing like this about the design of the physical form or DNA of any other creature.

The uniqueness of humanity identified by Dr. Collins and affirmed by the book of Genesis is the central point in dispute between materialists and all the enduring traditions of mankind. Ancient texts, biblical and nonbiblical, speak of man with veneration ("Man is but little lower than the angels," as Psalm 8:5 puts it). Even pagan traditions sense that human origins are rooted in the will of a deity, a fact that grants mankind special status. Evolutionists don't see things that way. They have taken a step backward from any position ever seen in human history. They believe that life literally comes from nothing and that human beings are the result of a series of accidents.

Their worldview offers no reason to speak of the dignity of man at all. Purely a material entity, no soul inspires him and no heaven awaits him. Born by sheer chance out of absolute nothingness, he enters a world without meaning that can offer him nothing except a return to that same empty darkness at whatever time and in whatever manner chance demands. No moral code can claim his obedience for there is no one who could write such a code and there is no reason he should pay attention to it if such a thing should be offered. Evolutionary man is guided only by the raw instincts of his animal nature. Within evolutionary thought there is no place for any of the dreams that have raised up civilization, dreams that gave us the renaissance and the enlightenment, the industrial revolution, and the age of information. There is only the bleak promise of the survival of the fittest celebrated in *Mein Kampf* by the human monster who thought he understood Darwin better than anyone else. In light of all this, would it kill us to listen to people like the director of the Human Genome Project and the many hundreds of other scientists who, in one way or another, have found technical fault with Darwinian evolution? Are the high school biology teachers right when they warn us that we will surely bring another dark age down upon history if we even for a moment entertain the hope that there is a God who loves us and who has crafted into our nature special gifts that give us dignity greater than the beast?

One and the Same

The text of Genesis makes one final and rather remarkable point about humanity. The text goes out of its way to tell us that humankind was created male and female. Genesis not only claims that the original code upon which the human being is based included within it a blueprint for both male and female expressions, but also speaks of two distinct steps in this encoding process: "So God created man in His own image; in the image of God He created him; male and female He created them" (Gen. 1:27).

The key words in this text are the pronouns. As usual with every description of an act of creation in Genesis, God reports the process of composing an "image" or mathematical blueprint. This blueprint is described as singular (a "he") and yet, it turns out that this "he" contains information necessary for the construction of both a "male and female" version of humanity. Then, surprisingly, at the end of verse 27 this same singular code is said to generate a "them," referring to two separate creatures, one distinctly male and the other female.

I believe the structure of this verse offers strong support for the interpretation of Genesis that I have offered throughout this book. The verse clearly treats the human creature as information, a code that contains great potential but a code that serves only as a blueprint for a creature as distinguished from the material expression of that creature. As pure mathematical code, the human being is described as both male and female at the same time. St. Paul was describing this original code when he wrote that in the perspective of our identity as information, that is in the eyes of Christ, "there is neither male nor female" (Gal. 3:28). The second reference in Genesis, however, seems to refer to something quite different. In the second part of the text the male and female components are separate, suggesting that what is described there is the physical expression of the governing mathematical code.

The implication is astonishing. This text tells us that the first physical object to appear in our universe was a human being. The rest of the material universe subsequently emerged when, according to the text of Genesis, man began to name the creatures of his world. We know that the generation of the physical universe was coincident with the formation of man because the text tells us that as soon as he sprang to life Adam began measuring and observing things: "Whatever Adam called each living creature, that was its name" (Gen. 2:19b).

The Genesis text here perfectly describes the process of observation and measurement by which, according to quantum physics, quantum energy fields are collapsed into specific material objects, a process physicist John Wheeler called "Genesis by Observership." It is important to note that in Genesis, the material adumbration of the mathematical possibilities encoded throughout the rest of chapter one only emerges coincidentally with the appearance of man the observer. In the text of Genesis it is only at this point in the narrative, in the same verse where human observation and measurement (naming) are first mentioned, that the elements of creation are reported to assume physical form: "Out of the ground the Lord God *formed* every beast of the field and every bird of the air, and brought them to Adam to see what he would call them" (Gen. 2:19a, emphasis added).

This verse directly connects the emergence of creatures as physical entities to the moment when Adam began naming those creatures. The first part of the verse tells us *what* God did (formed physical creatures); the second part tells us *how* He did it (through the agency of man). The message of Day Six in Genesis is clear. God determines the possibilities of existence but it is humankind that, by the way it observes and measures things, decides which of those possibilities will be realized in physical reality and which possibilities will remain, in this universe at least, an unrealized dream.

Chapter 10

Summing the Divine: The Seventh Day of Creation

Thus the heavens and the earth, and all the host of them were finished. And on the seventh day God ended his work which he had done, and He rested on the seventh day from all His work which He had done. Then God blessed the seventh day and sanctified it, because in it He rested from all His work which God had created and made.

—Genesis 2:1–3

After a journey through the six days of creation, it is obvious that the seventh day is different. It is not a day of creation at all; it is a day of rest. That the seventh is a day of rest testifies to the fact that it is seven earth days that concern us here. The Bible gives us no warrant to say that God's rest was a long vacation embracing vast ages. Genesis is describing a real rest on a real day.

The rest of God is important because it mandates a response by man. By resting, God establishes rest as a vital part of life on earth. If we refuse to observe at least one day of rest every week we are ignoring an example set by the Creator Himself and that would seem more than a little foolish. The fact that many do just that does nothing to change the foolhardiness of the behavior, it merely expands its scope.

There is an additional reason for us to observe a day of rest. In the Bible a day of rest is specifically enjoined as a celebration of the act of creation. The question facing us today is will we forget where we and our universe come from? As we have seen throughout this book, nothing good comes of forgetting God. No science can overcome a false premise, which means that if there is a God any results that are gained from research based on the set of working assumptions called *methodological atheism* will always be to some degree wrong. Should He exist, God would by necessity be a critical factor that would have to be taken into consideration in order to fully and accurately understand any phenomenon. In society in general it would seem difficult to sustain a moral code unless its principles were grounded in something or someone beyond our individual opinions. Not only is it hard to see why certain specific norms should be honored, it is hard to understand why there should be any rules at all if there is no lawgiver or judge. Psychologically, the breakdown of personal values and family life evident in our increasingly secularized culture makes the point that no human being thrives without a relationship to their Creator. It is important to remember that the sanctification of the Sabbath was not a part of the Jewish faith (which began at Mt. Sinai) nor is it tied to any one culture. The Sabbath was set apart for us on Day Seven of creation week, before religion or civilization began. This tells us that the act of honoring our Creator is not a religious issue; it is a fundamental component of human health and well-being.

The Sabbath day was a day of rest and therefore there is no Day Seven narrative. Genesis does, however, present a number, the number seven. As with the other days, the meaning of the seventh day can be found in the number. Moses said, "Teach us to number our days, that we may apply our hearts to wisdom" (Ps. 90:12). In that prayer the author of Genesis is, I believe, giving us the key to the interpretation of his book. Wisdom, that is, insight into the blueprint used to structure our universe, begins when we attend to the numbers of the days that appear in Genesis, including the number seven.

We can begin our study of the number seven by stating with assurance that this number is definitely sacred. In the Bible the number seven appears 463 times. Variants fill the text as well: "sevens" two times, "sevenfold" six times, "seventh" 120 times, "seventy" sixty-one times. In all, seven and variants of seven recur more than six hundred times in the Bible. But these totals do not reflect the impressive number of events, lists of names, miracle stories, wise sayings, symbols, and word patterns that are arranged in groups of seven. We will cite a few of these later in this chapter when we explore their meaning.

The question might then arise, if the number seven is so sacred and important is that number the mathematical expression of God just as six is the number of man? To this the answer must be no. God is not bound by any of the numbers we have been studying here. The numbers are a code structuring the informational content of creation. No number, not one or seven or anything between, defines God because God is the one who writes the numbers, He is not subject to them. To put that another way, God is not part of time or space. Isaiah makes this very clear when he writes of God, "It is he who inhabits eternity" (Isa. 57:15).

In the biblical worldview, eternity means something different than what it means in popular use. We know time as a series of days so it is natural for us to think of eternity as the same thing except longer. But biblical eternity is not extended time, it is a realm completely outside time. The main characteristic of eternity in the biblical usage is pattern. Eternity makes itself known in the material world through the iteration and reiteration of patterns. Eternity itself, that is, the actual patterns that are reflected in nature, is completely beyond time because the patterns themselves, being information, have no material quality. As the director of the German Federal Institute of Physics and Technology, Werner Gitt reminds us, "There is no known natural law through which matter can give rise to information, neither is any physical process or material phenomenon known that can do this."[1] Since eternity consists in patterns it is a different order of reality from anything we experience.

If seven is not the number of God what, then, is it? Seven is the presence of God within the reduced conditions of time/space. As we have seen, in Pythagorean thought the number seven was not treated as a product of other numbers, it possessed in itself an altogether unique mystical power. I suspect that this Pythagorean reverence for the number seven found its origin in the text of Genesis where that number is clearly associated with God. But while the number seven is connected with God in Genesis, the connection is narrow in scope and application. The Genesis reference to the number seven occurs in the context of the account of Creation and appears at the point where the act of creation has been completed. It is obvious that God's existence as a being with His own identity preceded Day Seven and every other day in Genesis. This means that seven is the number of God in the context of His relationship to His finished and inhabited creation. Seven is God but only in whatever form He may occasionally assume within the bubble-verse for purposes of communicating with the creatures there.

We need to underscore the limited significance of the number seven as applied to God. God is utterly beyond this bubble-verse. Everything in the universe we generate by our observations is always artificial and temporary because we have reduced it to material expression. This means that there will always be a difference between God as He is in Himself and God as we encounter Him in the physical world of our manufacture. The God we meet in the material world will be God redacted to fit the universe we have observed into existence.

God Made Known

Humanity is not able to invade the realm of God but God can make Himself known to us if He wishes to do so and that is what the number seven is all about. The number seven is the incarnation of God within the material universe. That incarnation has a name. The Bible tells us that the number seven deity who has been working behind the scenes throughout human history is none other than Jesus Christ. Christ is God's manifestation within the physical realm. St. John tells us: "No one has seen God at any time. The only begotten Son, who is in the bosom of the Father, He has declared Him" (John 1:18). Later in the same gospel this dialog is recorded:

> [Jesus says], "If you had known Me, you would have known My Father also; and from now on you know Him and have seen Him." Philip said to Him, "Lord, show us the Father, and it is sufficient for us." Jesus said to him, "Have I been with you so long, and yet you have not known Me, Philip? He who has seen Me has seen the Father; so how can you say, 'Show us the Father'?"
>
> —John 14:9

The core message of the Bible is that the Creator has entered the creation to save it. But entering the bubble-verse means that the one who enters becomes subject to its conditions: four dimensions not ten; informational waves full of unlimited potential that have been collapsed and reduced to particularized material expressions by mankind for purposes of measurement and analysis. For God to enter our world and our experience means that God must become man because there is no other way in which we could meet Him. We live in a material world. God as He is in his immaterial nature, what Luther called the "hidden God," is of no use to us. If God is to reach us and we are to touch God, He must become as we are. This is why God made the sacrifice that St. Paul celebrates:

> Let this mind be in you which was also in Christ Jesus, who, being in the form of God did not consider it robbery to be equal with God but made Himself of no reputation, taking the form of a bondservant, and coming in the likeness of men. And being found in appearance as a man, He humbled Himself and became obedient to the point of death, even the death of the cross.
>
> —Phil. 2:5–8

Unless you know some neat trick that will overrule the laws of physics, there is no other way for God to meet us or for us to encounter God. The humanity of Jesus is powerful confirmation of the concepts glimpsed by the ancients and now confirmed by quantum physics. Our world is the product of our observations and measurements, which reduce everything to a physical level. If God wants to enter a relationship with us, He must accept the limitations of physical existence. God's willingness to do this tells us how wide and how deep is the love of God for you and me that He would give up so much and accept such a profound loss to save us. Given the severe limitations our bubble-verse imposes on any possible contact between ourselves and God, it becomes easy to see why the Bible presents Jesus Christ as the absolutely unique and irreplaceable revelation of God. Once we understand how the universe works, we can appreciate why Jesus' intensely exclusivist statement recorded in the gospel of John simply has to be true: "I am the way, the truth, and the life. No one comes to the Father except through me" (John 14:6).

The basic points to remember are two ideas: the Creator of the universe is beyond enumeration, but when he meets us in our world He is the number seven and the number seven, wherever we encounter it in Genesis and throughout the Bible, is always Jesus Christ.

Spiritual Eternity

The limitations imposed by our artificial universe go far to clarify aspects of the biblical text that cause some people to dismiss the Bible as a collection of myths and legends. Throughout the Bible we read reports of encounters with angels, demons, and even Satan himself that are described in blatantly physical terms. These encounters involve appearances that are unquestionably anthropomorphic. Adding to their offensiveness for many people, the entities in question are usually seen by a limited number of people under localized and isolated conditions. Some readers, put off by these stories, join Judas in *Jesus Christ Superstar* in asking why God always seems to meet people in such backward places and strange times and doesn't just reveal Himself by way of some fabulous heavenly special effect.

In an effort to overcome the specificity of these stories, many try to redefine these encounters by imagining that they are allegories pointing to some deeper and more universal truth. What this approach fails to grasp is the fact that people can only meet the universal in the particular. Our universe exists naturally as information but we cannot make use of it in that form. We collapse the information to physical form so we can observe and measure it. Then, when we are done measuring and studying, we write a report, take a picture, or make a journal entry thereby restoring the experience to its status as information. But the physical encounter is a necessary part of this process. Until something is tied to a specific time and place and has taken on a visible shape, it really doesn't do us much good.

All the spiritual entities needed to satisfy the dreams of the wildest pagan cult could be circulating around us right now but it wouldn't mean anything to us because we wouldn't encounter them unless by observation and measurement we were permitted to collapse them to a physical particularity that would inevitably be different from their true nature. If we, then, reject reports of spiritual encounters because they seem too anthropomorphic, we will end up evicting all spiritual influences and entities from this planet. We should always remember that when we encounter a spiritual reality we do not meet it in its authentic form. The book of Jude, for example, reminds us that when angels came to earth in the time of Noah they took physical form and in the process fell prey to lust and so created the problem of hybridized humanity that forced God to destroy most of the gene pool in the event known as Noah's flood. These angels fell into sin and eventually died because, as Jude puts it, "They did not keep their proper domain but left their own abode" (Jude 1:6). They came determined to enter the world of human experience but to do that any spiritual entity will have to become as we are, physical beings whose lives are directed by appetites and limited by the qualities that our observations impose upon physical existence, including the inevitability of death. All real encounters with God and with godly things short of heaven will have a naively literal character not because God is naive and literal but because we are.

To illustrate the sort of problem this way of seeing things solves, let me offer a brief example involving the millennial rule of

Christ. We are told throughout the Bible that Christ will establish a kingdom on earth in fulfillment of many Old Testament promises to Israel. In this kingdom He will be joined by the saints who had been taken up into heaven but who will now reign with Him on earth for one thousand years. What troubles many is that the lives of these heavenly saints are described in limited terms. Even in the millennial kingdom imperfection will still exist and the saints will be subject to death (the Bible suggests that life spans will normally exceed one hundred years but people will not live forever). The question is, how can heavenly beings die?

The answer is that the bubble-verse is what it is and all who enter here become subject to its limitations. To be here is to be subject to observation including self-observation that involves a collapse into material form and to possess material form is to die. Even Jesus Himself was subject to death. The good news is that while our circumstances in this world, like Jesus' circumstances, are reduced in expression they are not reduced in essence. The information that we truly are persists unchanged (unless it is deleted and trashed and only God can do that). As I have said before, I like to think of our physical lives in computer terms. Coming into this world is like downloading information from the Internet. Downloading does not erase the source from which the information is drawn. This means that nothing that happens to us here can permanently harm or destroy us unless we lose our connection to the mind of God, where the essence of our being is safely stored as information. Death is a problem only within the framework of a materialist perspective, within which our aging and continually dying body is all that we ever are or will be. When, as believers, we view life from a heavenly/informational perspective, the body becomes for us what the Bible says it is: an earth suit (or as Paul liked to say, a "tent") we have temporarily put on for purposes of experiencing our existence in a physical way. The event of death where we put off that earth suit has lost its frightening sting for us because it means nothing more than a resumption of our true identity as information. This is why the Bible can make the kind of outlandish promises we read in these texts:

> But if a wicked man turns away from all his sins ... he will surely live; he will not die.
>
> —Ezek. 18:21

> Whoever lives and believes in me will never die.
>
> —John 11:26

> Do not marvel, my brethren, if the world hates you. We know that we have passed from death to life.
>
> —1 John 3:13–14a

The idea behind these kinds of texts is that when we have regained an accurate sense of who we are, the limitations of physical reality, including death, no longer define us. Those who identify themselves as information have no reason to believe that they will die because information does not die.

Provocative Patterns

I want to conclude this study with a look at a topic that has woven its way throughout this book: the topic of patterns. I think the brief overview I am about to offer might prove interesting and suggestive. I'm pretty sure it will be fun.

The fundamental claim of this book has been that Genesis teaches us that the universe we experience is made of information that God set in place as patterns before anything material came into existence. That information is now being collapsed into a material form by human beings for purposes of observation and measurement. But the essence of the universe and its eternal expression continues to reside as information in the form of patterns. Genesis focuses on those patterns, telling us how they were drawn up and giving us a sense of what they look like. By doing so it implies that if we study our material universe with attention to its informational content we can discern those archetypes. The archetypes will, in turn, give us a glimpse into the mind of God. But implied in this is a warning. None of this will happen if we insist that the material processes that we see point to no larger truth. If we believe that the material world is the ultimate expression of reality, we will pay a severe price. This is the point St. Paul made to his readers at the start of his letter to the Romans:

> For since the creation of the world His invisible attributes are clearly seen, being understood by the things that are made, even His eternal power [lit. archetypal patterns] and Godhead, so that they are without excuse because although they knew God, they did not glorify Him as God, nor were thankful, but became futile in their thoughts, and their foolish hearts were darkened. Professing to be wise, they became fools and changed the glory of the incorruptible God into an image made like corruptible man and birds and four-footed animals and creeping things.
>
> —Rom. 1:20–23

The question I want to address is this: is the Bible really insistent on this notion that there are patterns in nature that clearly reveal to any honest seeker the creative activity of God? We know the patterns are there. Chaos science has shown us the effect of the strange attractor. Quantum physics has identified preset energy levels that form stable mathematical arrangements that have no physical reality apart from the activity of an observer. The question I am posing is a matter of scriptural interpretation. Is this topic of patterns a basic *biblical* theme? And if it is, does the Bible confirm what we have said about the importance of these reiterated patterns? Does it support the idea that eternal patterns rather than their temporary physical manifestations are the essential expression of reality?

The theme of patterns is, in fact, pervasive throughout scripture, but to encounter this theme we must dig a little below the surface of the text. Those who study the Bible know that it is an unusual book. For one thing, it takes itself very seriously. Speaking through the pen of St. Paul, the Spirit that gave us the Bible says this: "All scripture is given by inspiration of God, and is profitable for doctrine, for reproof, for correction, for instruction in righteousness that the man of God may be complete, thoroughly equipped for every good work" (2 Tim. 3:16–17).

Note the words *all scripture*. This verse is referring to every single part of every book of the Bible and it is telling us that the entire span of the Bible is equally inspired. The question is, how deeply does the inspiration go? This question divides churches and believers. Some extend inspiration only to the general message of the Bible. They encourage readers to get the gist of the message and not worry about the details. The problem with this approach, a style of reading characteristic of liberal denominations, is that it invariably reduces the message of the Bible to a few commonplace observations that usually sound very much like things everyone already knows. One might have thought that the Creator of the universe would have something new and interesting to say.

Jesus took a very different attitude toward the inspiration of the Bible. He said of the book, "Not one jot, not one tittle shall pass from the law until all be fulfilled" (Matt. 5:18). Jots and tittles roughly correspond to crossing the t's and dotting the i's in English. In this quote, Jesus is saying that the inspiration of Scripture extends to the tiniest marks on each page. The implication is that there are no wasted words in the Bible. Every little thing is there for a reason.

When we remember that for every word in the Bible there is a reason that word is there, we have set the scene for having some fun and excitement while studying the book. By respecting every minute detail in the text including the arrangements of its words, we have created the possibility for seeing the patterns that might hide there. This is an important step because if it should turn out that the Bible is chock full of deliberately constructed patterns, that discovery would provide strong support for the notion advanced here that the Holy Spirit who guided the construction of the Bible really cares about patterns and is inviting us to search everywhere in the universe for the mind of God by looking for patterns.

Even the most casual Bible students know there are interesting patterns that present themselves in its text. For one simple example, the book is full of acrostics. Psalms 37, 111, 112, and 119 are arranged as acrostics following the order of the Hebrew alphabet.

The lengthy text of Psalm 119 stands out because it is elegantly organized into sections, each section beginning with a letter of the Hebrew alphabet in sequence, and each line in each section beginning with the letter that began that section. There are many easily accessible patterns of this kind and good Bible introductions will describe them.

A Book of Numbers

The most fascinating use of patterns in the Bible involves numbers. We have seen that seven is the archetype coding God incarnate in space/time (not God in Himself but God manifest in our material world). This means that seven is the number of Jesus Christ, who is the form of God become visible. We have also seen that the number seven fills the Scripture from beginning to end. But few people realize how pervasive this number really is.

The story of Jesus is a tale told by sevens. Some of these are well known. There are seven beatitudes in the Sermon on the Mount, seven last words spoken by Jesus on the cross, and seven petitions in the Lord's Prayer. The book of Revelation describes Jesus' return to Earth as an event replete with sevens, letters to seven churches, the breaking of seven seals, the sounding of seven trumpets, and the pouring of seven bowls (the list is almost endless). The future coming of the Messiah was symbolized in the Old Testament by seven articles of furniture in the Tabernacle, the seven times seven of the Jubilee year, the seven years it took Solomon to build the temple, and the seven festivals that filled the Jewish religious calendar. Again, the list of Messianic sevens could be extended to fill a book longer than this one. The reality is that every time an event occurs that manifests the presence of God on earth from the seven days of creation to the seven years of the final tribulation, the number seven is prominent.

I want to focus on one example that serves two purposes. My example demonstrates the close association between Jesus Christ and the number seven and includes an invitation from Jesus Himself to pay close attention to the numbers involved in what looks like a simple story. The story I refer to is the feeding of the five thousand. This is one of the very few events recalled in all four Gospels. Actually, there are two stories involved. In the gospels of Mark and Matthew the feeding of five thousand people is closely followed by an account of the feeding of four thousand. In this analysis I will focus on Mark's presentation of both stories.

The basic outlines of the accounts are widely known. Jesus and His disciples find themselves surrounded by huge crowds gathered in remote areas. Jesus is concerned that the people need to eat something. He proposes that His disciples feed the people, but the disciples object, pointing out that they can't afford to buy food. Jesus then inquires about any food that the disciples might have on hand.

This is the first point in the story where numbers enter the picture. It is important that Jesus does not ask if the disciples have food or what kind of food might be available. Jesus' question is specific. In each story he asks, "*How many* loaves do you have" (Mark 6:38; 8:5 emphasis added), a question designed to elicit a number. In each case the number that emerges is seven, although the content is different. One story involves a mix of five loaves and two fish while in the other story there are seven loaves. The menu for the proposed meal is not important here; what Jesus has asked for is a number and the number turns out to be seven.

As the story unfolds, Jesus blesses the food and it is distributed. Not only does everyone get enough to eat but there are basketfuls of leftovers. This is the second point at which numbers enter the story. In a surprising twist, the accounts each focus on exactly *how many* baskets were filled with leftovers. At the feeding of the five thousand we are told that the people took up twelve baskets (Mark 6:43). The feeding of the four thousand yielded seven large baskets of leftover fragments (Mark 8:8).

What is the meaning of these events? The key to grasping the message lies entirely in the numbers. Jesus Himself makes that point. Frustrated at the inability of His disciples to make sense of either sign, Jesus eventually explained the significance of these events this way:

> Do you not yet perceive nor understand? Is your heart still hardened? Having eyes do you not see? And having ears, do you not hear? And do you not remember? When I broke the five loaves for the five thousand, *how many* baskets full of fragments did you take up? They said to Him, "Twelve." Also, when I broke the seven for the four thousand, *how many* large baskets full of fragments did you take up? And they said, "Seven." So He said to them, "How is it you do not understand?"
>
> —Mark 8:19–21, emphasis added

Jesus might well be asking that question of us as well. Do we grasp the meaning of these events any better than the disciples? We start to move in the right direction when we note the words repeated twice in Jesus' question. Jesus does not ask "what did I say" or "what did you disciples do". He asks *how many.* The point of revelation in these stories lies in the numbers.

It is helpful to know that the two events happened in very different places. The feeding of the five thousand took place near the Sea of Galilee in Jewish territory. The feeding of the four thousand occurred in a region called the Decapolis, an area populated by Gentiles. Knowing that, interpretation of the stories requires nothing more than reading the numbers.

At the feeding of five thousand in Jewish territory, twelve baskets of leftovers were collected. As a combination formed by six added to six, twelve suggests the idea of one person ruling another person and as such is associated with governance. For example, Israel was founded by twelve patriarchs and was organized into twelve tribes. The early church was led by twelve apostles, a number deemed so important that when Judas, one of the twelve, committed suicide the rest of the disciples immediately drew lots to replace him. Twelve is the number of a person acting authoritatively within the framework of human history. The combination of the numbers twelve and seven with the land of Israel where the miracle took place tells us that the feeding of the five thousand was a prophetic event. The account looks forward to the establishment of Christ's rule on earth at the millennium, an event that fulfills God's distinctive promises to Israel. The feeding is a foretaste of the miraculous abundance that will characterize life in those days. It also makes it clear that the fulfillment of God's promises to Israel is connected to the presence of Jesus Christ, who will at that time rule Israel as its Messiah. The focus on the future of Israel is confirmed by the fact that five thousand were fed. The number five thousand is a heightened form of fifty which was the number of the year of Jubilee that came at the end of every forty-ninth year (7x7=49). The Jubilee year was an anticipation of the messianic age and was a reminder of the unique relationship enjoyed between God and His chosen people. It is also noteworthy that in this story Jesus makes the people sit down in groups of fifty and one hundred (a variant of fifty), highlighting again the Jubilee aspect of that event.

The story of the feeding of the four thousand, a story also replete with sevens but set in Gentile lands, is a prophetic anticipation of the coming age of the church. St. Paul spoke of the church as "the mystery which has been hidden from ages and from generations, but now has been revealed to His saints" (Col. 1:26). God's decision to extend to pagans an opportunity to participate in blessings promised only to Israel came as a complete surprise that found no hint in Old Testament prophecy. This second feeding was the first indication that God had prepared a gift for all the nations of the world. The fact that the focus number that appears in the text is the number seven tells us that this gift is entirely supernatural and is not a product of some sort of religious development. The church is not a product of history but an interruption of the flow of history interposed by the grace of God. The meaning of the event is underscored by the number four thousand. That number consists of a heightened form of the number 40 which is 5 x 8. Five, as we know, is the number of the chthonic, reflecting the slavery to instinct that binds the Gentile soul. The number eight, sounding as it does the beginning of a new octave in a scale, represents an event utterly outside the ordinary flow of history that is represented by the numbers one through seven. Eight is pure miracle, an appropriate archetype for the church, which begins in miracle at Pentecost and supernaturally departs from history at the rapture.

The point to grasp from this little Bible study is how important it is to attend to the numbers in examining a biblical text because it is in the numbers that the message is often revealed. What makes these stories unique is that Jesus clearly told us that their message lay in the numbers. But how many other texts are just like that even though no one warns us to pay attention to the math? How often do we glide over the numbers in a text or fail to count the number of participants in an event or names on a list and as a result miss the whole point of that passage? It is worth remembering that Jesus told us that "the hairs of your head are all *numbered*" (Matt. 10:30, emphasis added). In the same way, when God made his covenant with Abraham he illustrated the meaning of the event by inviting Abraham to go outside and said, "Look now toward heaven, and count the stars if you are able to *number* them" (Gen. 15:5, emphasis added). If God cares that much about numbers, maybe we should start counting things as well, at least when it comes to the text of the Bible.

The Sum of the Parts

The presence of patterns deliberately woven deep into the text of the Bible becomes even more obvious and the patterns take on more interesting forms when we turn to a type of analysis known as *gematria*. Hebrew, Latin, and Greek share a common characteristic in that they all use letters to express numbers (the reason why the Greek term *logos* referred to so often in this book renders the English notion of both number and word is because the two were inseparable). In Latin only a few letters were used as numbers (interestingly enough if you total the value of the six letters so used the result is the famous number 666). In Greek and Hebrew every letter serves as a number. This means that every word, sentence, and phrase in the Bible, which for the most part was written in Greek and Hebrew, has a distinct numerical value. Gematria refers to a way of studying ancient texts written in those languages that emphasizes the numerical sense of the letters as well as the semantics. When an effort is made to decipher a text in terms of its gematria, passages that might have seemed dry or even nonsensical can suddenly bloom with unexpected meaning. One ancient and intriguing Bible mystery provides a sample of this form of analysis.

In the Torah, the first five books of the Bible, we are told how God freed the people of Israel from slavery in Egypt and led them to the Promised Land. On their long journey they suffered many misadventures. One of the strangest is recounted in Numbers 21:5–9. The passage tells us that Israel grew tired of its long journey and rebelled against Moses, the leader God had appointed. Angry at their discontent, God sent "fiery serpents among the people" (Num. 21:6). Those bitten by the serpents died. As a result of the many deaths, the people came to Moses and repented and Moses agreed to pray for them. At this point the story takes its strange twist: "Then the Lord said to Moses, make a fiery serpent, and set it on a pole; and it shall be that everyone who is bitten, when he looks at it, shall live. So Moses made a bronze serpent, and put it on a pole; and so it was, if a serpent had bitten anyone, when he looked at the bronze serpent, he lived" (Num. 21:8).

That story is more than a little strange. The serpent is a symbol associated not with healing but with the devil. Bronze is also symbolic, being connected throughout the Bible with the brazenness of human sin. And yet here we are told that just to look upon this dreadful object raised up on a pole meant healing and new life for Israel.

Moses' serpent pole became in time a snare for the Hebrew people in part because it made no sense. When the temple was built in Jerusalem, the people raised this bizarre pole in that building, named it Nehushtan, and started to burn incense to it as to an idol (2 Kings 18:4). Eventually, King Hezekiah ordered the relic smashed and burned as part of a general reform of religious practices throughout Israel.

Today, Nehushtan stands as a reminder of how the Bible is an integrated message system in which each part must be understood in the context of the whole book. It is only in the third chapter of the gospel of Luke that the significance of this object becomes clear to a person reading through the Bible. In a conversation with the

Jewish leader Nicodemus, Jesus for the first time revealed in words the meaning of the serpent pole. He said to Nicodemus: "And as Moses lifted up the serpent in the wilderness, even so must the Son of Man be lifted up that whoever believes in Him should not perish but have eternal life" (John 3:14–15). The serpent pole was a prophetic symbol pointing forward to the cross of Jesus Christ and to the messianic work He accomplished there. The question is, could Israel have known this before Jesus revealed it? The answer to that question is yes, but the revelation lay not in the words of any text but in the numbers.

There is an odd correspondence between the numerical expressions of certain names that appear in the Bible, a correspondence that, to some people, may seem disturbing. In Hebrew the English word *serpent* is rendered as *nakhash*. The gematria or numerical value of that word is 358. The important Hebrew name *Mashiyakh*, which is rendered in English as *Messiah*, has the same numerical value of 358. To those who could read the text of the Bible in the language of mathematics, the connection between a coming savior and the apparently satanic figure of a serpent was never unknown.

This mathematical correspondence between serpent and Christ is not restricted to the single tale of Nehushtan. The name *Jesus of Nazareth* rendered in Greek expresses the number 2197. The Greek name rendering Satan also expresses the number 2197. The Hebrew name for Christ, translated "His Annointed," expresses the number 364, which is also the number of the usual Hebrew name for Satan (*Ha-Shatan*). There are many similar matches that could be cited. In the realm of gematria, the connection between Christ and Satan is persistent and deep.

Have we discovered here a secret that Dan Brown could turn into another bestseller like *The DaVinci Code*? The fact is, there is no secret here at all. In the book of Isaiah we are told something about the coming Messiah that people who are unaware of the Bible's interest in mathematical patterns might miss. Speaking of the suffering servant of God, the prophet tells us, "He was numbered with the transgressors" (Isa. 53:12). Only someone who knows that every single letter in the Hebrew and Greek alphabet is also a number can understand what Isaiah was telling us. Isaiah was hinting that interpretation using gematria could provide clues that would allow a careful reader to identify the Messiah and even recognize His name when he appeared. Isaiah then went on to describe the work of the Messiah in a way that made the message in the numbers explicit by saying that "He bore the sins of many" (Isa. 53:12). Hundreds of years later, St. Paul underscored this same theme when, speaking of the death of Jesus Christ, he wrote, "Christ has redeemed us from the curse of the law *having become a curse for us* (for it is written 'Cursed is everyone who hangs on a tree')" (Gal. 3:13, emphasis added).

Jesus Christ redeemed us by taking to Himself the full weight of all the sins of humanity, past, present, and future. By doing so He became the embodiment of sin itself. This is why when He was ordered to defend Himself by Pilate and the High Priest he stood silent. He could offer no defense because He did, in fact, stand guilty—of our sins. Paul made this even clearer when he observed that, "[God] made Him who knew no sin to *be sin* for us, that we might become the righteousness of God in Him" (2 Cor. 5:21, emphasis added). Jesus won our victory by going to the cross and becoming the same as Satan, the very embodiment of moral darkness. When He was subsequently raised from the dead, this meant that His sacrifice had been accepted and freedom from slavery to chthonic forces had been secured for all who believed.

The thing to notice here is that the entire gospel story existed as a mathematical pattern before it took narrative shape. Throughout the Old Testament era, an extraordinary revelation had awaited anyone who studied the Bible in such a way as to find that pattern. This would seem to suggest that similar revelations await those who examine the universe not with an eye to describing its pedestrian objects but with an intention to discern its deep patterns.

Messianic Message

We conclude this search for patterns in the Bible by returning to the realm of semantics. Bestsellers have recently filled bookshelves with fantastic examples of revelations gained by applying the principles of something called *equidistant letter sequencing* to the study of the Bible. Thanks to computer technology, researchers have been able to decipher potential messages that are formed by letters separated by an equal number of spaces and letters in the biblical text. To be considered meaningful, a message generated by equidistant patterns normally should be related to the content of the text as encountered in ordinary reading.

An easy example that can help us grasp how equidistant letter sequencing works occurs in the first five books of the Bible. These are the most revered books of the Hebrew canon and together bear the name *Torah*. If the beginning verses of Genesis and Exodus are subjected to this kind of analysis, the word *Torah* appears beginning with the first tau (T) of each book and continuing with each subsequent letter appearing at exactly forty-nine-letter intervals. The number 49 is 7x7, a number full of profound messianic implications; it is also the number of the important year of Jubilee that occurred at the end of every forty-nine years. No such pattern appears in the third book of the Torah, the book of Leviticus. When the last two books of the Torah, Numbers and Deuteronomy, are subjected to this form of analysis, the word *Torah* appears again at the same numerical intervals but this time written backward. This yields a result that can be presented as follows:

TORAH TORAH [LEVITICUS] HAROT HAROT

When the first Yod in the book of Leviticus is subjected to this same kind of analysis, a word is revealed at seven letter intervals but that word is not *Torah*, it is *YHWH*, the name of God. There may be a message in this. All of the Torah is to be considered an integrated whole and all of it points to one thing: the presence of God among His people Israel.

This simple example reminds us once again that the Bible is an integrated message system in which everything is there for a reason and nothing is trivial. The rabbis had a saying that when the Messiah comes He will interpret everything in the Bible, including the empty spaces. Jesus may well have done that very thing in a variety of ways we do not fully appreciate today. For example, when He said, "Many who are first will be last, and the last first" (Matt. 19:30), He may have been hinting at a secret even more mysterious than a grand reversal of expectations at the judgment that will end history. He may have been pointing to hidden messages, what the rabbis call *remes*, that can only be grasped by reading certain texts both forward and backward. Concerning the possibility of such hidden messages, the great rabbi Eliyahu ben Shlomo (1720–1797) wrote:

> All that was, is, and will be unto the end of time is included in the Torah, the first five books of the Bible ... and not merely in a general sense but including the details of every person individually and the most minute details of everything that happened to him from the day of his birth until his death likewise of every kind of animal and beast and living thing that exists, and of herbage, and of all that grows or is inert.[2]

The Bible speaks of a "Book of Life" in which every name and every event is recorded. Some have speculated that this book may actually be the Bible itself, read in accordance with procedures even more arcane than equidistant lettering sequences. Confirmation that this radical proposition might in some sense be true has emerged from scientific study of the Bible. A careful study of the book of Genesis was completed by a group of statisticians and reported in the refereed journal *Statistical Science*. Using equidistant letter sequencing, these researchers found within that book the names of sixty-six famous rabbis. Along with these rabbis' names, they also found listed the dates of the births or deaths of each man. They determined that the chance such a pattern could be a matter of coincidence was less than one in 2.5 billion.[3]

Other studies in which prophetic texts that seem relevant to current events in the Middle East yielded detailed information about protagonists directly involved with those events when subjected to equidistant letter sequencing have been widely reported. Just as widely reported has been the fact that similar results have been obtained by subjecting *Moby Dick* and other books to equidistant sequencing.

Because of reports like that, many have dismissed the entire search for literary patterns of this sort as a kind of pseudo-science. I believe this is a fundamental mistake. The very fact that equidistant letter sequencing yields results when applied to any literary text makes the point of this chapter. The existence of such sequences is not important because they suggest that there is something mysterious about the Bible. The sequences are important because they imply that our lives and our world may indeed be shaped by patterns. The fact that the Bible reveals the reality of patterns underscores the thesis of this book that information that finds its source somewhere beyond our space/time domain is the force that structures life in our universe. The fact that patterns appear in various forms of literature as well as in both nature and human culture does not in any way challenge that thesis; it confirms it.

The universe as we know it is the expression of a mathematical code. That code is the software of life. Everything that occurs happens within the limitations established by that code. Quantum physics tells us that human beings can, by their observations, briefly collapse the code into the artificial form we call material existence for purposes of measurement and analysis. But that same physics also tells us that the permanent existence of the universe takes the form of the information coded into our universe from the beginning. The act of encoding our universe is the subject of the book of Genesis. The work of collapsing the code into material expressions is the ongoing work of humanity.

The point of Genesis and the point of this book is that we will not do our job well nor will we really understand our universe unless we grasp the limited but important place of man in the cosmic scheme and see the world we inhabit for what it is. That world is in its source true and holy but in its physical expression it is only a useful illusion. We will not break through the illusion by studying the appearances of things as materialists hope to do. To understand ourselves or our universe we must direct our attention to the mysteries of God. The great King Solomon, the wisest man who ever lived, had it exactly right when he penned the words that end this book:

> The fear of the Lord is the beginning of wisdom.
>
> —Prov 1:7

Note carefully the number of that text.

Endnotes

Chapter Two

1. Charles Darwin. *The Origin of Species By Means of Natural Selection or the Preservation of the Favored Races in the Struggle for Life,* (London: John Murray, 1859/1872), 281–282.
2. Niles Eldridge. *Reinventing Darwin,* (New York: John Wiley and Sons 1995), 156.
3. Ibid., 70.
4. Ibid., 69.
5. Darwin, 316–317.
6. Private correspondence quoted in Eldridge, 13.
7. Private correspondence quoted in William Irvine, *Apes, Angels and Victorians* (New York: Time Incorporated. 1955), 176.
8. Cited in Rupert Sheldrake, *The Presence of the Past* (Rochester, Vermont: Park Street Press. 1995), 281.
9. Burton Guttman, *Biology*. 6th ed. (New York: McGraw-Hill. 1999) 470. quoted in Jonathan Wells. *Icons of Evolution*. (Washington D.C.: Regnery Publishing, 2000), 189.
10. Peter H. Raven and George B. Johnson, *Biology* 5th ed. (Boston: WCB/McGraw-Hill. 1999), 334. Quoted in Wells, 185.

11. Michael Behe, *Darwin's Black Box: The Biochemical Challenge to Evolution,* (New York: Free Press, 1996), 39.
12. Dr. Colin Patterson, Senior Paleontologist, British Museum of Natural History, London. Transcript of discussion at the American Museum of Natural History, New York City, November 5, 1981, accessed at www.arn.org/arnproducts/audios/co10. Htm.
13. Irvine, 176.
14. Wells, 102.
15. Richard Dawkins, *The Blind Watchmaker: Why the Theory of Evolution Reveals a Universe Without Design* (New York: WW. Norton Company, 1996). 384. quoted in Werner Gitt, *In the Beginning Was Information* (Bielefeld, Germany:Christliche Literatur-Verbreitung e.V., 2000), 136.
16. Wells, 203.
17. Phillip Johnson, *The Wedge of Truth,* (Downers Grove, Illinois: InterVarsity Press, 1999), 140.
18. William Demski, *Intelligent Design,* (Downers Grove, Illinois: InterVarsity Press, 1999), 125.
19. Dawkins, *The Blind Watchmaker,* 1, quoted in Demski, 125.
20. Dawkins, *The Blind Watchmaker,* Preface, quoted in Wells, 205.
21. Dawkins, *The Blind Watchmaker,* 683, quoted in Sheldrake, 281.
22. Dawkins, *The Blind Watchmaker,* 6.
23. Pew Forum on Religion and Public Life, July 7–17, 2005, pewformum.org.
24. Study reported on August 11, 2006, at www.Foxnews.com.
25. Quoted by Jonathan Witt in "Random Act of Design," *Touchstone Magazine,* 19, no. 8 (October 2006), The HCD Research Inc. letter is available at www.hcdi.net/polls/j5776.
26. Demski, 251.

27. Stephen Jay Gould, "Is a New and General Theory of Evolution Emerging?" *Paleobiology* 6, No.1 (1980): 119–130, quoted in Johnson, 71.
28. Rob Kaplan, "Coming of Age in the Origin of Species, 1880," quoted in *Science Says: A Collection of Quotations on the History, Meaning, and Practice of Science,* (New York: W.H. Freeman and Company, 2001), 13.
29. Paul Davies, "Bit Before It?," *New Scientist Magazine,* Issue 2171 (January 30, 1999) accessed at www.newscientist.com/article/mg161217710.100—bit-before-it. html.

Chapter Three

1. Werner Gitt, *In the Beginning was Information* (Bielefeld, Germany: Christliche Literatur-Verbreitung e. V. 2000), 19.
2. Ibid., 46.
3. K. Dose, *Tagungsbericht uber den ISSOL—Kongreb in Mainz* vom 10. Bis 15. Juli 1983, quoted in Gitt, 107.
4. Philolaus. *Fragment 4,* quoted in Charles Kahn, *Pythagoras and the Pythagoreans* Indianapolis, Indiana: Hackett Publishing. 2001, 25.
5. Aristotle, *Metaphysics* 1080b18, cited in Arthur Fairbanks, ed. and trans., *The First Philosophers of Greece* (London: K. Paul, Trench, Trubner. 1898; (Hanover Historical Texts Project scanned and proofread for Internet publication by Aaron Gulyas, 1998).
6. Aristotle, *Metaphysics* I5; 985 b23-986b8, quoted in Fairbanks.
7. Aristotle, *Physics* IV.6, 213b22, quoted in Kahn, 29.
8. Numenus, *Fragment 11,* quoted in Kahn, 123.
9. Hermes *I.G.P. 16* and *Dox 655,* quoted in Fairbanks.
10. Galileo quoted in Rupert Sheldrake, *The Presence of the Past: Morphic Resonance and the Habits of Nature* (Rochester Vermont: Park Street Press.1995), 23–24.

11. Sheldrake, 22.
12. Philo, *De opificio Mundi* 90, quoted in Kahn, 101.
13. James Ussher, The Annals of the World (Green Forest, Arkansas: Master Books. 2006), 121.
14. Ibid.
15. Numenus, *Fragment 8,* quoted in Kahn 119.
16. Johannes Kepler, *Harmonice Mundi* Book III, Chapter 1, quoted in Kahn, 162.

Chapter Four

1. Augustine of Hippo. *De Genese ad Litteram,* www.ediacara.org/august.htmls.
2. Amit Goswami, *The Self-aware Universe* (New York: Jeremy P. Tarcher, 1995), 59.
3. Ibid., 59.
4. Michio Kaku, *Parallel Worlds: A Journey Through Creation, Higher Dimensions, and the Future of the Cosmos* (New York: Doubleday. 2005), 153.
5. John Gribbon, *In Search of Schrodinger's Cat* (New York: Bantam Books, 1984), 173.
6. Goswami, 59–60.
7. David Darling, *Equations of Eternity* (New York: Hyperion. 1993), 102.
8. Kaku, 350.
9. Kaku, 153.
10. Banesh Hoffman, *The Strange Story of the Quantum* (Magnolia. Ma. 1963), 217 quoted in Gribbon, 192.
11. Darling, 102.
12. Goswami, 141.
13. Gribbon, 79.

14. Werner Gitt, *In the Beginning was Information* (Bielefeld, Germany: Christliche literatur-Verbreitung, 2000), 31.
15. Gitt, 79.
16. Ernst Cassier, *Language and Myth* Translated by Suzanna K. Langer. (New York: Harper and Brothers, 1946), 8 quoted in Arthur Zajonc, *Catching the Light.* (New York: Bantam Books. 1993), 61.
17. Gitt, 46.
18. Kahn, *Pythagoras and the Pythagoreans,* 29.
19. Ibid., 54.

Chapter Five

1. Jeans quoted in Michio Kaku, *Hyperspace* (New York: Oxford University Press, 1994), 116.
2. *Introductio Arithmeticae* 1.4.2 quoted in Kahn, 113.
3. Goswami, 98.
4. Arthur Zajonc, *Catching the Light.* (New York: Bantam Books, 1993), 265.
5. Chuck Missler, *Cosmic Codes, Hidden Messages from the Edge of Eternity* (Coeur d' Alene, Idaho: Koinonia House, 1998), 124.
6. Goswami, 190.

Chapter Six

1. Kahn, 57.

Chapter Seven

1. Albert Einstein, Letter of Condolence to the family of his friend, Michelle Besso, March 21, 1955. Quoted in Missler, 39.
2. Kaku, 166.
3. Sir James Jeans, *The Mysterious Universe.* (Cambridge, Cambridge University Press, 1930) quoted in Missler, 279.

4. Robert Nadeau and Menas Kafatos, *The Non-local Universe: The New Physics and Matters of the Mind.* (New York: Oxford University Press, 1999), 154.

Chapter Eight

1. *Introduction to Arithmetic* 1.6.1 quoted in Kahn, 114.
2. Gribbon, 172.
3. Zajonc, 340.

Chapter Nine

1. Kaku, 171–172.
2. Darling, 102.
3. Jean B. De Lamarck, *Philosophie Zoologique* Vol. 1 (Paris, 1809), 104f quoted in Gitt, 81.
4. Alfred North Whitehead, *Science and the Modern World.* (New York: The Macmillan Company, 1929) 80, quoted in Richard Milton, *Alternative Science.* (Rochester, Vermont: Park Street Press 1996), 220.
5. Zajonic, 204.
6. Galileo Galilei, *The Assayer in Discovery and Opinions,* quoted in Zajonoic, 76.
7. Aristotle, *On the Heavens*, quoted in Kaplan, 69.
8. Fritjof Capra, *The Tao of Physics: An Exploration of the Parallels Between Modern Physics and Eastern Mysticism* (Berkeley, California: Shambala Publishing, 2000) 18.
9. Missler, 41.
10. Gitt, 46.
11. Ibid., 81.
12. Kaku, 171.
13. Milton, 74.
14. Quoted by Jonathan Witt in "Random Act of Design," *Touchstone Magazine,* 19, no. 8 (October 2006), Internet archive.

Chapter Ten

1. Gitt, 79.
2. Missler, 124.
3. Ibid., 139.

Printed in the United States
132295LV00009B/15/P

9 781414 111605